Research between Science, Society and Politics

Research between Science, Society and Politics

The History and Scientific Development of Green Chemistry

Johan Alfredo Linthorst

Eburon
Utrecht 2023

The financial support of laboratory supplier Boom B.V. (Meppel, the Netherlands) for the purpose of printing this book is gratefully acknowledged.

ISBN 978-94-6301-434-2

Eburon Academic Publishers, Utrecht, The Netherlands
www.eburon.nl

Voor mijn lieve vrouw, Elza.

Table of Contents

Acknowledgements

After my graduation in 2004 at the University of Groningen, I decided to start a career as a fulltime chemistry teacher at a secondary school. In the same year I also started to do research in the history of chemistry. However, with one leg in education and one leg in science my rewarding journey as an "external PhD candidate" was at times difficult to organize. Fortunately, people around me did understand. I therefore wish to thank in particular the organizations and people mentioned below for their support during my long PhD journey.

Because during my PhD journey the geographical distance between my home and Maastricht University ranged from 250 km to 8.000 km, visits to Maastricht were rare. Meetings with my supervisor in Utrecht sometimes offered a good alternative.

My supervisor, Ernst Homburg, believed in this research from the start, which gave me the stamina to overcome many barriers and struggles. Ernst, I am extremely thankful for your continual support, criticism and honesty. I have truly benefitted from our conversations and your knowledge of the history of chemistry, as well as from your critical comments during the writing process of this thesis. I have appreciated all your input and enthusiasm and I am looking forward to continue sharing our joint interests in the history of chemistry.

I am also grateful to the Royal Society of Chemistry, the American Chemical Society, the Science History Institute and the Royal Netherlands Chemical Society for the hospitality during my study of their archives. My research trips to the archives in London, Philadelphia, PA, and Washington, DC, were financially supported by the Society for the History of Alchemy and Chemistry (New Scholars Award), the Science History Institute (travel grant), the Royal Holland Society of Sciences and Humanities (Pieter Langerhuizen Stipendium) and the Descartes Centre for the History and Philosophy of the Sciences and the Humanities of Utrecht University (2009 fellowship). I am very grateful for their support.

The Descartes Centre for the History and Philosophy of the Sciences and the Humanities also provided me with library services and ICT-infrastructure to carry out my research. Special words of thanks go to Bert Theunissen and also to Lodewijk Palm who collected and sent research documents to Curaçao.

I would also like to thank the interviewees who took the time and had the patience to provide me with valuable insights on the subject in the course of my research.

The work to carry out this research was not always easy. But I received help from several persons. I would like to mention my colleagues at CSG Dingstede for their support: Johanna, Marielle, Paul, Dick (all chemistry colleagues), and Hannie (management). I would like to thank my colleague Arjen Boerstra in particular, and also Jan Hop, my former chemistry teacher. Arjen and Jan, as you both know, our concerns about the Dutch education system reached the headlines of Dutch newspapers many times and received the attention in the heart of Dutch democracy. Next to that, our joint efforts gave me the pleasure to continue this research.

For a good balance in life over the years friends and family have been very important for me. Dear Jan, Roelof, Bert, Erik (all "Matties" from my native village, Ruinen), Henk, Arnold, Jarno, Heidi and many more, I thank you all for the drinks we had together. Peter, in spite of your devotion to Feyenoord, our relationship is very special due to our shared commitment to stand up for the rights of the poorest people on Curaçao. Our efforts made a difference, and sharpened my mind.

There has been no bigger thing in Dutch tradition than "Sinterklaas" to me, because in that time of the year I received many poems, full of jokes, which reminded me to finish this PhD thesis. Geert, Roelie, Marko, Marina, Kor, Frank, my sister Marga, my brother Harrie, Martha and of course Jacelien in particular: thank you very much for that support. Harrie, as you know, you are always on my mind and in my heart. My father, Jan, and my mother, Aafje (1939-2013), who sadly passed away due to Parkinson disease, are also thanked for their support and for giving me the opportunity to start academic education. This is something which I will never forget.

All these dear friends and family are surpassed by my sons and my wife. Dear Aaron and Ivo, I love you both to the moon and back and I promise you to spend more time together, on Curaçao for instance. Last but not least, I thank the woman of my life. Elsie, what and where would I be without you? This book is also your book.

List of Abbreviations

ACS	American Chemical Society
C_1Chemistry	Koolwaterstoffenconversie [Conversion of hydrocarbons] (NL)
C_3	Stichting Communicatie Centrum Chemie [Foundation Communication Centre Chemistry] (NL)
CAA	Clean Air Act, 1970 (USA)
CEFIC	Conseil Européen des Fédérations de l'Industrie Chimique [European Chemical Industry Council]
CEI	Committee on Environmental Improvement (ACS)
CFCs	Chlorofluorocarbons
COCI	Committee on Chemistry and Industry (IUPAC)
CPA	Control of Pollution Act, 1974 (UK)
CS	Chemical Society (UK)
DCMR	Dienst Centraal Milieubeheer Rijnmond [Central Environmental Protection Agency for the Rijnmond area] (NL)
DCO	Stichting Duurzame Chemie Ontwikkeling [Foundation for Sustainable Chemistry Development] (NL)
DEC	Division of Environmental Chemistry (ACS)
DETR	Department of Environment, Transport and the Regions (UK)
DfE	Design for the Environment program (EPA)
DoE	Department of the Environment (UK)
DTO	Duurzame Technologische Ontwikkeling programma [program for Sustainable Technological Development] (NL)
DTO-Chemie	Duurzame Technologische Ontwikkeling Chemie programma [program for Sustainable Technological Development Chemistry] (NL)
EC	European Community
ECG	Environmental Chemistry Group (RSC)
EEC	European Economic Community
EHSC	Environment, Health and Safety Committee (RSC)
EPA	Environmental Protection Agency (USA)
EPrA	Environmental Protection Act, 1990 (UK)
EPSRC	Engineering and Physical Sciences Research Council (UK)

EPTS	Environmental Protection Technology Scheme (UK)
EuCheMS	European Association for Chemical and Molecular Sciences (now EuChemS – European Chemical Society)
GCI	Green Chemistry Institute (ACS)
GCN	Green Chemistry Network (RSC)
GDCh	Gesellschaft Deutscher Chemiker [the German Chemical Society]
GDP	Gross Domestic Product
GPC	Geïntegreerde plantconversie [Whole crop utilization] (NL)
HMIP	Her Majesty's Inspectorate of Pollution (UK)
IM	Inspectie Milieuhygiëne [Inspectorate of Environmental Health] (NL)
IPC	Integrated Pollution Control (UK)
IUPAC	International Union of Pure and Applied Chemistry
KNCV	Koninklijke Nederlandse Chemische Vereniging [Royal Netherlands Chemical Society]
LRTAP	Long-Range Transboundary Air Pollution (Convention on, 1979)
MC	Sectie Milieuchemie [Division Environmental Chemistry] (KNCV)
NEPA	National Environmental Policy Act, 1969 (USA)
NEPP	Nationaal Milieubeleidsplan [National Environmental Policy Plan] (NL)
NERC	Natural Environment Research Council (UK)
NO_x	Nitric oxide and nitrogen dioxide
NOTA	Nederlandse Organisatie voor Technologisch Aspectenonderzoek [Netherlands Organization for Technology Assessment]
NSF	National Science Foundation (USA)
NWO	Nederlandse Organisatie voor Wetenschappelijk Onderzoek [The Dutch Research Council]
OECD	Organisation for Economic Co-operation and Development
OPPT	Office of Pollution Prevention and Toxics (EPA)
OSTP	Office of Science and Technology Policy (White House) (USA)
PCBs	Polychlorinated biphenyls
PFC	Fijnchemische procestechnologie [Fine chemicals process technology] (NL)
PPA	Pollution Prevention Act, 1990 (USA)
PPO	Pollution Prevention Office (EPA)

PZC	Fotovoltaïsche zonne-energie [Photovoltaic solar energy] (NL)
PRISMA	Project Industriële Successen Met Afvalpreventie [Project Industrial Successes with Waste Prevention] (NL)
RCEP	Royal Commission on Environmental Pollution (UK)
RCRA	Resource Conservation and Recovery Act (USA)
RIC	Royal Institute of Chemistry (UK)
RIVM	Rijksinstituut voor Volksgezondheid en Milieu [National Institute for Public Health and the Environment] (NL)
RSAS	Royal Swedish Academy of Sciences
RSC	Royal Society of Chemistry (UK)
SAR	Structure-activity relationships
SPR	Specialist Periodical Reports (UK)
STD	Sustainable Technological Development program (NL)
STD-Chemistry	Sustainable Technological Development Chemistry program (NL)
STS	Science and Technology Studies
TF	Technology Foresight program (UK)
TSCA	Toxic Substances Control Act (USA)
UN	United Nations
UNEP	United Nations Environment Program
VNCI	Vereniging van de Nederlandse Chemische Industrie [Association of the Dutch Chemical Industry]
VOCs	Volatile organic compounds
VROM	Ministerie van Volkshuisvesting, Ruimtelijke Ordening en Milieubeheer [Ministry for Housing, Physical Planning and Environment] (NL)
VVC	Vezelversterkte composiet-materialen [Fiber-reinforced composite materials] (NL)
WWII	World War II

Chapter 1: Introduction: the Emergence of Green Chemistry

1.1 Chemistry and the Environment

During large parts of its history chemistry was seen as a science of progress, admired by many. Since a few decades though, public debates on chemistry have been quite critical, mainly for environmental reasons. Often "chemistry" and "the environment" are seen as opposing concepts. Seen in that light is it interesting to note that in the course of the 1990s a new field of chemistry emerged, which has been called "Green Chemistry". The birth, evolution and scientific meaning of that new field are the subject of this book.

The first decades after World War II were a period of strong focus on economic growth; but societies paid a negative price too. Rachel Carson's famous book *Silent Spring* documented the serious consequences of chemicals such as pesticides with respect to ecological damage and public health.[1] Societies faced environmental catastrophes with chemical factories, e.g., the Givaudan-La Roche plant at Seveso (1976), and the Union Carbide plant at Bhopal (1984).[2] They had also to cope with growing amounts of chemical waste. As a result, chemical industries were seriously criticized by the public in many Western countries.

The public environmental concerns received international political recognition at the 1972 United Nations Conference on the Human Environment in Stockholm.[3] Chemicals also caught headlines due to the negative impact of chlorofluorocarbons (CFCs) on the ozone layer, as well as emissions of sulfur dioxide to the atmosphere that caused transboundary air pollution and acid rain.[4] As a consequence, new environmental policies and measures were introduced in modern democracies that were often based on geopolitical agreements, e.g., the phasing out of CFCs (1987), the policy goal of sustainable development (1987) and the agreement on the reduction of greenhouse gas emissions (1992).[5]

---

[1] Rachel Carson, *Silent Spring* (New York, NY: Houghton Mifflin Company, 1962).
[2] The journal *Ambix* devoted a special issue to this topic in March 2002.
[3] United Nations, "Declaration of the United Nations Conference on the Human Environment," in *Report of the United Nations Conference on the Human Environment* (Stockholm: United Nations, *1972*), 3-5.
[4] Anthony N. Penna, *The Human Footprint: A Global Environmental History* (Chichester: Wiley-Blackwell, 2010).
[5] The years of the geopolitical agreements are between brackets. United Nations, *Montreal Protocol on Substances that Deplete the Ozone Layer* (Montreal: United Nations, 1987); United Nations World Commission on Environment and Development, *Our Common Future* (Oxford: Oxford University

As a result, the public meaning attached to chemistry and to the chemical industry became ambiguous. Countries experienced major economic benefits from chemical industries in terms of employment and contributions to their national income. Chemical industries, and as a result, also societies, relied on robust knowledge of the chemical science, which significantly shaped our material culture. But people were also worried about environmental issues, which led to a growing demand for chemical knowledge concerning the environment.[6] These developments also strongly influenced the organizational structure of professional and learned bodies such as chemical societies. In most industrial countries, divisions, working groups and committees on so-called "environmental chemistry" were established within those societies during the early 1970s. Environmental chemistry emerged as a subfield that contributed to the analysis and solving of environmental problems.[7] During those years also new journals in that field were launched, such as *Toxicological & Environmental Chemistry Reviews* in 1972.[8]

In this thesis I investigate the emergence and further development of scientific subfields of the chemical sciences devoted to environmental and ecological issues: first environmental chemistry, and then the rather different field of green, or sustainable, chemistry. I do this by focusing on three highly developed industrial countries – the USA, the UK, and the Netherlands – and within those countries I look in particular to the developments within the national chemical societies, and how these differed, or did not differ, from one country to the other. This historical study of environmental and green chemistry aims to improve our understanding of the responses by chemists and chemical societies to public environmental concerns, and of changing knowledge demands in modern democracies.

The paths along which environmental and green chemistry emerged in the USA, the UK and the Netherlands are analyzed in this thesis through the lens of the history of the chemical societies studied: the American Chemical Society (ACS), the Royal Society of Chemistry (RSC) and the Royal Netherlands Chemical Society [Koninklijke Nederlandse

---

Press, 1987); United Nations, *United Nations Framework Convention on Climate Change* (Rio de Janeiro: United Nations, 1992).

6 Hannah Gay, "Before and After *Silent Spring*: from Chemical Pesticides to Biological Control and Integrated Pest Management-Britain, 1945-1980," *Ambix* 59 (2012): 88-108.

7 Throughout this thesis I follow a commonly used terminology by historians of science: discipline (e.g., chemistry and physics), subfield (e.g., organic chemistry) and scientific specialty (e.g., bio-organic photochemistry). For further reading see: Eduard Jan Rinia, "Measurement and Evaluation of Interdisciplinary Research and Knowledge Transfer," (PhD diss., Leiden University, 2007), 20. In Leiden University Repository; available from https://openaccess.leidenuniv.nl/handle/1887/9923, accessed May 22, 2022.

8 This journal is currently titled *Toxicological & Environmental Chemistry*.

Chemische Vereniging] (KNCV). These routes will be explained in a context of changing public environmental values and of differences in political cultures and styles of policy making concerning science and the environment. These changes in environmental values and policies influenced the three chemical societies in different ways. A comparative analysis of the ACS, the RSC and the KNCV helps to better understand the construction and communication of chemical knowledge with respect to environmental problems in those three nations, leading to three different trajectories in the emergence of "green chemistry".

The story of environmental and green chemistry presented in this book can also illustrate the changing societal context of the chemical sciences in the late 1980s. During those years people were increasingly worried about the quality of the environment. This increase of public environmental concerns ran parallel to the worldwide attention paid to the goal of sustainable development. At the same time, new demands were formulated for chemical knowledge in order to prevent pollution at the source. But, as I will argue in this book, chemical societies and other actors in the USA, the UK and the Netherlands also shaped their own science and environmental policies in that context. They did that in different ways. I will demonstrate how culturally conditioned differences influenced the emergence and further development of a new movement and scientific field that holds, even today, great promises for society: green chemistry.

Green chemistry emerged during the early 1990s at the cutting-edge of chemical science and society, and has since then witnessed a huge growth in both academic and industrial circles. At the same time government policies concerning science and the environment became increasingly democratized.[9] In order to understand the origins of green chemistry I also take that factor into account.

From the earliest publications on green chemistry in scientific journals there has been much discussion about its precise (scientific) meaning. Walter Leitner, for instance, argued in 1999 that clean chemistry, sustainable chemistry and green chemistry essentially do have the same meaning.[10] But members of the Working Party on Green and Sustainable Chemistry (European Association for Chemical and Molecular Sciences) argued in 2008 that there was still 'no consensus in the scientific community' on the prefixes "green" and "sustainable".[11]

---

[9] See for example: Esther Turnhout et al., "New Roles of Science in Society: Different Repertoires of Knowledge Brokering," *Science and Public Policy* 40 (2013): 354-365.
[10] Walter Leitner, "Toward Benign Ends," *Science* 284 (1999): 1780-1781.
[11] P. Tundo, "Report of the 1st meeting of the Working party on Green and Sustainable Chemistry" (European Association for Chemical and Molecular Sciences, Rome, February 28, 2008). See also in: Johan A. Linthorst, "An Overview: Origins and Development of Green Chemistry," *Foundations of*

More recently, (former) editors of *Green Chemistry* commented on the growth and identity of green chemistry, and in particular on the question which green chemistry labeled research rightly counted as green chemistry proper, and which did not.[12] The identity and status of green chemistry thus appears to be a matter of dispute, and as a result its promises and meaning for society are a bit obscure too. Against that background this thesis aims to clarify these disputes and analyze the evolving meaning and importance of green chemistry for science and society by tracing its scientific development. It will address in particular its origins in the USA, the UK and the Netherlands, both from an internal scientific and from a socio-cultural point of view. My analysis in this book therefore aims to be of interest to historians of science, policy makers, and chemists as well.

Societies have provided infrastructures and funding for chemical research. Major policy activities and lobbying have been undertaken in the 1990s by the ACS, the RSC and the KNCV in order to secure the legitimacy of the chemical sciences, in view of the deteriorating public image of chemistry.[13] ACS President S. Allen Heininger argued in the early 1990s that 'improving the public's perception of chemistry' was a key aim of the ACS, because the 'American public is highly interested in environmental issues, and a common misconception is that chemists, chemistry, chemicals, and the chemical industry are all detriments to the environment'.[14] So, the negative public image of chemistry should certainly be taken into account to understand the emergence and recognition of green chemistry.

Chemistry 12 (2010): 55-68. See also: "Who is EuChemS," The European Chemical Society, accessed May 22, 2022, EuChemS Who is EuChemS – EuChemS.

[12] James H. Clark et al., "15 Years of Green Chemistry," *Green Chemistry* 16 (2014): 18-23. See also: Carlos A. Marques and Adélio A.S.C. Machado, "Environmental Sustainability: Implications and Limitations to Green Chemistry," *Foundations of Chemistry* 16 (2014): 125-147.

[13] Johan A. Linthorst, "The Image of Chemistry and Curriculum Changes," *Educacion Quimica* 23 (2012): 240-242; Ernst Homburg and Elisabeth Vaupel, "Conclusion," in *Hazardous Chemicals: Agents of Risk and Change, 1800-2000*, ed. Ernst Homburg and Elisabeth Vaupel (New York: Berghahn Books, 2019), 382-386. See for chemical societies for instance: ACS, *American Chemical Society Annual Report 1996* (Washington, DC: ACS, 1997), 14; RSC, *The Royal Society of Chemistry: Annual Review 1995* (London: RSC, 1996), 15; KNCV, "KNCV Jaarverslag 1990," *Chemisch Weekblad* 87 (1991): 196. See also: Michelle Francl, "How to Counteract Chemophobia," *Nature Chemistry* 5 (2013): 439-440; Carla Morais, "Storytelling with Chemistry and Related Hands-on activities: Informal Learning Experiences to Prevent "Chemophobia" and Promote Young Children's Scientific Literacy," *Journal of Chemical Education* 92 (2015): 58-65; Pierre Laszlo, "On the Self-image of Chemists, 1950-2000," *Hyle* 12 (2006): 99-130; Peter H. Spitz, "The Chemical Industry and the Environment: Meeting the Challenge," in *The Chemical Industry at the Millennium: Maturity, Restructuring, and Globalization*, ed. Peter H. Spitz (Philadelphia, PA: Chemical Heritage Press, 2003), 206-246.

[14] S. Allen Heininger, "President's Statement," in ACS, *American Chemical Society Annual Report 1991* (Washington, DC: ACS, 1992), 3.

1.2 Studies on the Development of Environmental and Green Chemistry

Only a limited number of studies have been conducted that systematically examined the history of environmental chemistry and green chemistry. The major studies that do exist will be discussed below.

The history of environmental chemistry

There are many histories of chemistry, but most of these do not take the debates on chemistry and the environment on board. William Brock and Trevor Levere are exceptions. They emphasize the important impacts on the discipline by accidents in the chemical industry and instances of worldwide pollution, by for instance pesticides.[15] Moreover, specific historical studies on the emergence of environmental chemistry as a subfield of chemistry are scarce, and they were often written by environmental chemists themselves.

Environmental chemistry emerged in the early 1970s. The KNCV Division Environmental Chemistry [Sectie Milieuchemie] was established in 1970, and the ACS Division of Environmental Chemistry (DEC) in 1973.[16] In the UK, the Environment Group was founded in 1972 within the Industrial Division of the Chemical Society (CS), one of the predecessors of the RSC.[17] In the same period research and educational programs on environmental chemistry were launched at universities in the USA, the UK and the Netherlands.[18] In 1974, the environmental chemist G. Fred Lee argued in a special issue on

---

[15] William H. Brock, *The Fontana History of Chemistry* (London: Fontana Press, 1992), 656-662; Trevor H. Levere, *Transforming Matter: a History of Chemistry from Alchemy to the Buckyball* (Baltimore, MD: The Johns Hopkins University Press, 2001), 192-200.

[16] Stuurgroep Sectie Milieuchemie, *Koninklijke Nederlandse Chemische Vereniging: Sectie Milieuchemie*, August 4, 1970; KNCV, "Jaarverslag van de Koninklijke Nederlandse Chemische Vereniging over 1970," *Chemisch Weekblad* July 23/30 (1971): 28; DEC, "Division of Water, Air, and Waste Chemistry," April 10, 1973.

[17] Geoff Dickes, "A Brief History of the Environmental Chemistry Group," *Environmental Chemistry Group Newsletter* 1 (1995): 4-5; David H. Whiffen and Donald H. Hey, *The Royal Society of Chemistry: The First 150 Years* (London: RSC, 1991), 206-209.

[18] See for the USA: Glen E. Gordon, "The Role of Chemistry Departments in the Environmental Education of Chemists," *Journal of Chemical Education* 51 (1974): 769; George L.B. Pratt, "Educational Needs of Chemists in Pollution Control and Abatement," *Journal of Chemical Education* 51 (1974): 770-771; G. Fred Lee, "Graduate Education in Environmental Chemistry," *Journal of Chemical Education* 51 (1974): 772-774; D.D. Davis, "An Operational Graduate Program in Environmental Chemistry," *Journal of Chemical Education* 51 (1974): 775-776. See for the UK: Brendan Keely, "Environmental Chemistry at the University of York," *Environmental Chemistry Group Newsletter* 5 (1997): 3-4. See for the Netherlands: Werkgroep Aspekten van

environmental chemistry education of the *Journal of Chemical Education*, that this emergence in the USA was supported by 'substantial research funds'.[19] Environmental chemists who later reflected on their own subfield, identified the early 1970s as the period when environmental chemistry took off.[20] But the limits of the field were a matter of dispute. Donald Crosby, for instance, argued that the identity of environmental chemistry as a scientific subfield was hard to characterize due to the fact that the creation of knowledge in environmental chemistry is subject to a wide range of 'things, conditions or influences'.[21] The Dutch environmental chemist Jan Willem Copius Peereboom agreed with that, but emphasized that environmental chemistry was not only a subfield of chemistry, but also a 'specialism' within the interdisciplinary environmental sciences, which were based on both the social and the natural sciences.[22] He associated this chemical subfield thus also with social concerns. The environmental historian Christopher Hamlin positions environmental chemistry pretty close to actions of the people as well in his definition of the 'perilously vague term' environmental chemistry:[23]

> (…) a field of study: inquiry into the chemical systems of the environment, a concern with the effects of the environment on human health and the effects of human activities on environmental systems, (…)

The environmental historian Henk van Zon does the same, but also points to the impact of natural effects on environmental systems.[24]

Milieuproblematiek in de Chemische Opleiding, *De Milieuchemie in het Wetenschappelijk Onderwijs* (Den Haag: KNCV, 1976).

[19] Lee, "Graduate Education in Environmental Chemistry," 772. This special issue was USA oriented. See for the increases in research funding also: Laszlo, "On the Self-image of Chemists, 1950-2000," 114; Gay, "Before and After *Silent Spring*," 97-100.

[20] Dickes, "A Brief History," 4; Donald G. Crosby, "Review: Environmental Chemistry: an Overview," *Environmental Toxicology and Chemistry* 1 (1982): 1; J.W. Copius Peereboom and K. Bouwer, "Environmental Science 'Milieukunde' in the Netherlands: a Review," *The Science of the Total Environment* 129 (1993): 161.

[21] Crosby, "Review: Environmental Chemistry," 8.

[22] J.W. Copius Peereboom, *Chemie, Mens en Milieu: Schadelijke Stoffen in Milieu en Voeding, een Studie over Chemische Milieuverontreiniging* (Assen: Van Gorcum, 1976), VIII; Werkgroep Aspekten van Milieuproblematiek in de Chemische Opleiding, *De Milieuchemie in het Wetenschappelijk Onderwijs*, 4; Copius Peereboom and Bouwer, "Environmental Science 'Milieukunde' in the Netherlands," 157-158.

[23] Christopher Hamlin, "Between Knowledge and Action: Themes in the History of Environmental Chemistry," in *Chemical Sciences in the Modern World*, ed. Seymour H. Mauskopf (Philadelphia, PA: University of Pennsylvania Press, 1993), 295.

[24] Henk van Zon, "Geschiedenis van Milieuproblemen in Vogelvlucht," in *Milieuchemie. I. Onderzoeksmethoden in de Milieuchemie*, ed. M.A.M. Meester (Heerlen: Open Universiteit, 2000),

Despite these ambiguities, expectations were high: Crosby even predicted that environmental chemistry would become 'the chemistry of the 21st century'.[25] But despite those expectations, the emergence and growth of environmental chemistry from the early 1970s onwards has not specifically been studied by professional historians of chemistry.

I conclude that a comprehensive historical analysis of environmental chemistry is still lacking. Despite that, I also conclude that several authors recognized major influences by public environmental concerns on the identity of environmental chemistry. Crosby and Hamlin signaled difficulties in assessing the scientific character of environmental chemistry. In order to better understand the identity of this chemical subfield I will analyze in the next three chapters its emergence and development by studying in more depth the histories of the relevant divisions, or sections, of the three national chemical societies.

Studies on green chemistry and its history

Contrary to the history of environmental chemistry, the history of "green chemistry" has been studied in some depth by a number of authors. The term "green chemistry" entered the chemical sciences in the early 1990s, but by the turn of the century the use of this new term in chemical journals increased suddenly. That phenomenon became therefore subject of scholarly interest in the field of Science and Technology Studies (STS), which gained momentum in the mid-2000s and was often spurred by the laudable promise of green chemistry for society: sustainability. In order to understand the emergence and growth of green chemistry this scholarly interest led to different characterizations of green chemistry, which I have grouped in four different categories: (1) green chemistry as a scientific movement; (2) green chemistry as a social movement; (3) green chemistry as mere rhetoric; and (4) green chemistry as a (meta-)specialization within the chemical sciences.

Green chemistry as a scientific movement

One of the first STS-scholars who studied green chemistry thoroughly was Jody Roberts (Chemical Heritage Foundation), who in 2005 wrote a dissertation on the emergence of green

---

31-32 and 46-47. The following environmental chemists put forward similar characterizations: Dickes, "A Brief History," 4; Crosby, "Review: Environmental Chemistry," 1; Copius Peereboom and Bouwer, "Environmental Science 'Milieukunde' in the Netherlands," 161.

[25] Crosby, "Review: Environmental Chemistry," 6.

chemistry in the USA in the years 1990 to 2005.[26] He argues that the nature of environmental legislation changed in the USA with the passage of the Pollution Prevention Act (PPA) of 1990. Since then the focus on pollution prevention increased, with positive effects on both the environment and the economy.[27] The PPA also led to the formation of a green chemistry program within the US Environmental Protection Agency (EPA). According to Roberts, PPA was therefore a major turning point for the emergence of green chemistry. The roots of green chemistry in the USA are thus within the EPA, and chemists within that organization developed a program that received major political support in 1995, when the Presidential Green Chemistry Challenge Awards program was launched. One of its aims was to show society the benefits of the chemical sciences and chemical industry. The Awards also stimulated the further development of green chemistry.[28]

Roberts, who interviewed several early key-figures within the field of green chemistry, argues that these proponents were strategically creating a green chemistry movement.[29] In order to persuade other chemists to join the green chemistry community they created a historical narrative of chemistry, the environment and the economy.[30] In order to solve USA's environmental and economic problems, the logical outcome of that story was a unique selling-point for the chemical sciences: "green chemistry". In 1998, this 'alternative form of chemistry' was intellectually shaped through the establishment of the so-called "twelve principles" by two leading promoters of green chemistry: Paul Anastas and John Warner (both from the USA).[31]

These principles play a fundamental role in Roberts' characterization of green chemistry as a "scientific movement". He uses in his analysis the general theory of Scott Frickel and Neil Gross on the social dynamics and conditions of scientific movements, in which they included insights of various scholars within STS. These "scientific movements" are 'collective efforts to pursue research programs or projects for thought in the face of

---

[26] Jody A. Roberts, "Creating Green Chemistry: Discursive Strategies of a Scientific Movement," (PhD diss., Virginia Polytechnic Institute and State University, 2005), abstract. In VirginiaTechWorks; available from https://vtechworks.lib.vt.edu/handle/10919/27529, accessed May 22, 2022. The current name of the Chemical Heritage Foundation is Science History Institute (Philadelphia, PA).
[27] Ibid., 10-13.
[28] Ibid., 64-67 and 130.
[29] Roberts also attended green chemistry conferences.
[30] Roberts, "Creating Green Chemistry," 38-67.
[31] Ibid., abstract and 69-96. In 1998, Warner and Anastas published these principles in their handbook: Paul T. Anastas and John C. Warner, *Green Chemistry: Theory and Practice* (New York, NY: Oxford University Press, 1998), 29-55.

resistance from others in the scientific or intellectual community' and 'involve dramatic breaks with past practices'.[32] Roberts concludes:[33]

> I argue that we have little choice but to call it a scientific movement. (...) Additionally, I think that perhaps the problem lies not so much in calling this a movement, but in calling it scientific.

Roberts doubts in this characterization is mainly based on his views on the nature of the twelve principles of green chemistry and on its institutional origins. With respect to the latter, he argues that green chemistry initially emerged 'not out of any academic unit, let alone a prestigious one. And the leaders of the movement hail from a variety of institutions, but none from a particular place of prominence'.[34] This lack of intellectual leadership could thus be a reason for not calling green chemistry a scientific movement, according to Roberts. But, similar to dominant advocates of green chemistry in the USA, he also states:[35]

> The establishment of the definition and the [twelve] principles, I argue, constitutes an important move in constituting the field as a very specific interdisciplinary group with a forged identity and the beginnings of a system for determining what properly 'counts' as green chemistry.

Roberts acknowledges the work of his STS-colleagues Edward Woodhouse and Steve Breyman (see below) 'to address and overcome the social and political obstacles' in the recognition of green chemistry, but he also emphasizes the pivotal role of the twelve principles in giving green chemistry a *scientific* identity.[36] For that reasons Roberts characterizes green chemistry as a scientific movement, which aims at transforming chemistry

---

[32] Scott Frickel and Neil Gross, "A General Theory of Scientific/Intellectual Movements," *American Sociological Review* 70 (2005): 206-207.
[33] Roberts, "Creating Green Chemistry," 149.
[34] Ibid., 150.
[35] Ibid., abstract. See for instance the advocates of green chemistry: Paul T. Anastas and Mary M. Kirchhoff, "Origins, Current Status, and Future Challenges of Green Chemistry," *Accounts of Chemical Research* 35 (2002): 686-687.
[36] Roberts, "Creating Green Chemistry," 95. See also: Laura Maxim, "More than a Scientific Movement: Socio-political Influences on Green Chemistry Research in the United States and France," *Science & Technology Studies* 31 (2018): 24-46. Maxim also characterizes green chemistry as a scientific movement on the basis of Frickel and Gross.

into a green and sustainable practice on the basis of an interdisciplinary approach.[37] In order to achieve that goal, he predicted in 2005:[38]

> The [twelve] principles are also exclusionary, eliminating the participation of many through the strict technical focus of the principles. And finally, adherence and exaltation of the principles leads to unnecessary – and potentially dangerous – solidification of green chemistry as something other than chemistry in its broader sense.

The quotation shows that Roberts has doubts about the further growth of green chemistry, because it might stagnate due to the fact that the twelve principles impose limitations to green chemistry as a scientific movement. As a result, he argues that green chemistry will not accomplish what it sets out to do, that is to transform chemistry as a whole into a sustainable practice.[39]

Roberts wrote an unique and comprehensive historical study on the development of green chemistry, but mainly limited to the USA. Other in-depth studies on the emergence and characterization of green chemistry often take other viewpoints, as we will see.

Green chemistry as a social movement

Roberts referred in his work to an influential article by Woodhouse and Breyman from 2005.[40] These authors examine green chemistry as a social movement, which they defined as:[41]

> (…) people mobilizing themselves and other resources to change other people, social structures, and social relations using means which may be unconventional for ends which may include the radical.

---

[37] An interdisciplinary character is also noted recently by other STS scholars. See for instance: Leonardo Victor Marcelino, Adilson Luiz Pinto and Carlos Alberto Marques, "Intellectual Authorities and Hubs of Green Chemistry," *Transinformação* 32 (2020): e200031; Carlos Alberto Marques and Adélio A.S.C. Machado, "An Integrated Vision of the Green Chemistry Evolution along 25 Years," *Foundations of Chemistry* 23 (2021): 299-328.
[38] Roberts, "Creating Green Chemistry," 96.
[39] Ibid., abstract.
[40] Edward J. Woodhouse and Steve Breyman, "Green Chemistry as Social Movement?," *Science, Technology & Human Values* 30 (2005): 199-222.
[41] Ibid., 209.

Woodhouse and Breyman noted that only a minority of the chemists is concerned with green chemistry though. They therefore advocate for a much greater influence of the public on the development of green chemistry, which in their view is 'a new way of doing chemistry'.[42] In their analysis of the identity of green chemistry they observe the mixed use of different terms that all have an environmental connotation, e.g., green chemistry, clean chemistry, sustainable chemistry and environmentally benign chemistry.[43] Moreover, Woodhouse and Breyman claim that green chemistry strives to 'making chemicals and their production inherently benign'.[44] Based on their focus on elite actors they characterize green chemists, who are sharing the same ideologies and strategies, as follows:[45]

> Green chemists are united in seeing themselves as advancing the values of health, safety, and sustainability – and decidedly do not perceive themselves as participants in chemical industry "greenwashing".

Similar to Roberts, they contend that green chemists are not aiming for replacing chemical practices, but transforming them for achieving a major societal objective: sustainability.

Green chemistry as mere rhetoric

Alastair Iles, who builds on the influential article by Woodhouse and Breyman, characterized green chemistry in 2011 as a social movement too, which started in the USA and expanded to Europe.[46] But he also argued that some chemists consider green chemistry as a rhetorical tool 'to express their ecological concerns'.[47] The impact of rhetoric on the emergence of green chemistry, as well as the rhetoric of green chemistry, was studied by several STS scholars. For instance, Andrew Nelson and his coauthors in 2014 recognize an 'unclear and changing

---

[42] Ibid., 199. Iles also argues that green chemistry is missing societal input. See: Alastair Iles, "Greening Chemistry: Emerging Epistemic Political Tensions in California and the United States," *Public Understanding of Science* 22 (2011): 460-461.
[43] Roberts also identified a mixing of environmentally conscious terms. See: Roberts, "Creating Green Chemistry," 72-74. See also: Pietro Tundo and Fabio Aricò, "Green Chemistry on the Rise: Thoughts on the Short History of the Field," *Chemistry International* 29 (2007): 4.
[44] Woodhouse and Breyman, "Green Chemistry as Social Movement?," 205.
[45] Ibid., 200 and 210.
[46] Iles, "Greening Chemistry: Emerging Epistemic Political Tensions," 460.
[47] Ibid., 467.

status of green chemistry'.[48] Based on a statistical research methodology, they attempt to determine to what extent publications on green chemistry correspond to accepted green chemistry practices. They report 'suggestive evidence of extensive "greenwashing"' in a large set of journal articles that contain the term green chemistry.[49] They also argue that green chemistry labeling and proper environmentally friendly practices do often not go hand in hand.

In her reflection on the use of buzzwords in science policy, the French historian and philosopher of chemistry Bernadette Bensaude-Vincent in 2014 also discussed the emergence of the term green chemistry in science communication. She presents this as the logical outcome of the tension between the chemical sciences and the critical public that blames chemistry for its polluting operations. That led to a so-called crisis in society, according to Bensaude-Vincent, because the public proved insecure with respect to health and environmental risks involved. She argues:[50]

> The buzzword "green chemistry" also appeared as a safe and reasonable line of action against the background of intense conflicts. (…) Buzzwords are prescriptive but not imperative. They peacefully recruit people through more subtle ways, as they spread from mouth to mouth, from paper to paper, from institution to institution. Their performances increase as the number of people who catch them and include them in their own agendas increases. The scope of this power may therefore be assessed through the number/frequency of occurrences for a set/given period of time.

A (meta-)specialization within the chemical sciences?

Next to the analysis of rhetoric and theories on social and scientific movements discussed above, various scholars attempted to characterize the (internal) scientific identity and rise of green chemistry through the lens of scientific specialization.

For instance, Kidwai and Mohan chose in 2005 an anecdotal approach in order to characterize the internal scientific identity of green chemistry. They consider the rise of green

---

[48] Andrew Nelson et al., "Do Innovation Measures Actually Measure Innovation? Obliteration, Symbolic Adoption, and Other Finicky Challenges in Tracking Innovation Diffusion," *Research Policy* 43 (2014): 931.
[49] Ibid., 927.
[50] Bernadette Bensaude-Vincent, "The Politics of Buzzwords at the Interface of Technoscience, Market and Society: the Case of 'Public Engagement in Science'," *Public Understanding of Science* 23 (2014): 247. See also: Maxim, "More than a Scientific Movement," 40.

chemistry as a scientific 'revolution'.[51] In their view green chemistry is 'essentially a way of thinking' that 'utilizes a set of principles that seek to reduce the environmental impact of chemical processes'.[52] Kidwai and Mohan rely on the scientific power of the "twelve principles", which serve as 'guidelines for practicing chemists in developing and assessing how green a synthesis, compound, process or technology is'.[53] They argue in the journal *Foundations of Chemistry*:[54]

> Green Chemistry is not different from traditional chemistry in as much as it embraces the same creativity and innovation that has always been central to classical chemistry. However, there lies a difference in that historically synthetic chemists have not been seen to rank the environment very high in their priorities. (...) Since the types of chemicals and the types of transformations are much varied, so are the green chemistry solutions that have been proposed.

Kidwai and Mohan claim that green chemistry is highly varied in order to tackle societal concerns. Subsequently, they report about dozens of organic syntheses that demonstrate the fruitfulness of insights from green chemistry in order to produce less waste. In a recent paper, Mark Murphy lists in the same journal a number of chemical syntheses too, but he also mentions the intellectual roots of green chemistry in industry that emerged in advance of the 1990s, and which were steered by 'competitive and economic market needs'.[55]

The "twelve principles" have a pivotal role, because they function as rules in order to make chemistry more sustainable, according to Marianna Epicoco et al. in 2014.[56] In their view green chemists form a so-called epistemic community, which means that:[57]

---

[51] M. Kidwai and R. Mohan, "Green Chemistry: an Innovative Technology," *Foundations of Chemistry* 7 (2005): 269.
[52] Ibid., 285. See also: Astrid E. Schwarz, "Green Dreams of Reason. Green Nanotechnology Between Visions of Excess and Control," *NanoEthics* 3 (2009): 109-118; Anastas and Kirchhoff, "Origins, Current Status," 690.
[53] Kidwai and Mohan, "Green Chemistry," 271.
[54] Ibid.
[55] Mark A. Murphy, "Early Industrial Roots of Green Chemistry and the History of the BHC Ibuprofen Process Invention and its Quality Connection," *Foundations of Chemistry* 20 (2018): 121. For example, in the Netherlands green chemistry became also tied with chemical industry. See: Wetenschappelijke en Technologische Commissie voor de Biobased Economy, *Naar Groene Chemie en Groene Materialen: Kennis- en Innovatieagenda voor de Biobased Economy* (Den Haag: WTC, 2011).
[56] Marianna Epicoco, Vanessa Oltra and Maïder Saint Jean, "Knowledge Dynamics and Sources of Eco-innovation: Mapping the Green Chemistry Community," *Technological Forecasting & Social Change* 81 (2014): 391. This study is based on the analysis of citation networks and algorithms.
[57] Ibid.

> (…) epistemic communities produce knowledge as much as they set to influence politics. Epistemic communities emerge from a policy demand and the policy receptivity is a crucial issue for them: they have to produce "usable knowledge" and the knowledge they produce has to provide solutions in response to specific problems. (…) In the academic sphere, epistemic communities are constituted around groups of researchers who share a common cognitive objective of knowledge creation in a common research field.

Epicoco et al. thus attribute an applied research component to green chemistry, which is steered by policy demands. The research is carried out by chemists who are cognitively united, and is supported by newly developed networks. Intellectual and practical solutions are crucial in order to receive recognition and to survive as an epistemic community, according to these authors.

In comparison with Kidwai, Mohan and Epicoco et al., Jesper Sjöström advocates in 2016, together with Ingo Eilks and Vânia Zuin, for a different approach in order to characterize green chemistry as a kind of scientific specialization. Although he realizes that different characterizations are possible for green chemistry ranging from a 'movement' to a 'subdiscipline', Sjöström argues that green chemistry is a 'meta-discipline'.[58] He claims that since the 1960s the public image of chemistry deteriorated due to environmental pollution caused by chemical production. He states:[59]

> As a response to the "green challenge" and as a way to improve the image of chemistry, the so-called green chemistry movement started in the US in the early 1990s. (…) Green chemistry is not a scientific discipline in the normal sense. Instead it is a *meta-discipline* that covers most of chemistry and chemical engineering.

Sjöström thus argues that decreasing pollution and improving the public image were major drivers for the growth of green chemistry.[60] Moreover, he calls green chemistry a meta-

---

[58] Jesper Sjöström, Ingo Eilks and Vânia G. Zuin, "Towards Eco-reflexive Science Education: a Critical Reflection about Educational Implications of Green Chemistry," *Science & Education* 25 (2016): 322.
[59] Jesper Sjöström, "Green Chemistry in Perspective – Models for GC Activities and GC Policy and Knowledge Areas," *Green Chemistry* 8 (2006): 130.
[60] Maxim concludes the same for France. See: Laura Maxim, "Chemists' Responsibility for the Health Impacts of Chemicals: Green Chemistry and Its Relation to Toxicology," *Hyle* 23 (2017): 61-80; Maxim, "More than a Scientific Movement," 30. See for a philosophical discussion on the relationship between the environment and chemistry also: Joachim Schummer, "The Notion of Nature in Chemistry," *Studies in History and Philosophy of Science Part A* 34 (2003): 726. Schummer describes case studies that cover different centuries.

discipline and advocates the importance of 'social perspectives on different green chemistry activities and policy/knowledge areas' instead of a focus on the twelve principles.[61] According to Sjöström, a meta-discipline refers to:[62]

> Finally, integrating meta-perspectives (philosophical, historical, and socio-cultural) in the chemical research and education practice would not only provide chemical *Bildung* to practitioners; it could also improve the currently too rationalist and reductionist public image of chemistry, and in the long term create a more reflective discourse of chemistry.

Sjöström therefore employs a broad view on what green chemistry encompasses, which are 'all activities and policy/knowledge areas with the aim of greening of chemistry and chemical practice'.[63]

William Lynch, and Jean-Pierre Llored and Stéphane Sarrade also incorporate contextual values in characterizing green chemistry, but by applying moral or ethical perspectives in particular.[64] For instance, Lynch argues:[65]

> Green chemistry presents a constructive model for how ethical implications of engineering practice can be incorporated into an epistemological project of revisioning methods and standards in the field to remedy the normalization of deviance.

Lynch means with 'normalization of deviance' the fact that environmental problems are often easily accepted as inevitable. He relies on the scientific power of green chemistry in order to shape and transform chemical practices more environmentally friendly though. Llored and Sarrade also rely on that power and state that 'chemists must become capable of integrating the long-run consequences of their actions and productions'.[66] Hence they reflect on the development of green chemistry:[67]

---

[61] Sjöström, "Green Chemistry in Perspective," 135.
[62] Jesper Sjöström, "The Discourse of Chemistry (and Beyond)," *Hyle* 13 (2007): 94.
[63] Sjöström, "Green Chemistry in Perspective," 130.
[64] Sjöström would call this meta-perspectives.
[65] William T. Lynch, "Second-guessing Scientists and Engineers: Post Hoc Criticism and the Reform of Practice in Green Chemistry and Engineering," *Science and Engineering Ethics* 21 (2015): 1237.
[66] Jean-Pierre Llored and Stéphane Sarrade, "Connecting the Philosophy of Chemistry, Green Chemistry and Moral Philosophy," *Foundations of Chemistry* 18 (2016): 135.
[67] Ibid., 138.

> Green chemistry is currently in progress. It may succeed in reshaping and transforming chemistry or, maybe, it could fail. Its capacity to bring about a radical change depends on future consensus between policy makers, chemical practitioners and citizens in order to give green chemistry the power it rightly deserves.

They thus argue that green chemistry still 'could fail'. After more than twenty years of existence, resulting in many publications in scientific journals, it is still 'in progress'. The success of green chemistry ultimately depends on its power to change chemistry, in their view, and I will analyze that power and change further in this thesis.

1.3 Research Questions and Theoretical Framework

STS scholars thus used different (theoretical) perspectives in order to characterize the identity and growth of green chemistry. These perspectives emphasize different aspects, and interpretations, of green chemistry, but do not radically exclude each other. Movements, for instance, can be both scientific and social; and both movements and scientific specializations make use of rhetorical strategies.

Several authors emphasize that green chemistry was mainly a response to the negative public image of chemistry. Examples are Roberts, Iles, Woodhouse and Breyman, and Sjöström.[68] They focus mainly on the development of green chemistry in the USA. Chemical sciences are international in scope and specific developments vary from one political culture to another, as I will demonstrate in this thesis. For a deeper understanding of the emergence and growth of green chemistry it is important to analyze the development of research in green chemistry in both the USA and Europe, in relation to the different (risk) cultures in those political systems. This thesis therefore presents a comparative analysis of the developments in three different countries in order to answer the central research question: how did environmental concerns and the public image of chemistry shape the emergence and (scientific) development of green chemistry in the USA, the UK and the Netherlands?[69]

The authors discussed above employed different research methodologies, ranging from statistics, e.g., Nelson et al., to philosophical perspectives, e.g., Sjöström , Llored and Sarrade. The present study is mainly a historical analysis of the roles of three national chemical

---

[68] Roberts addresses the public image of chemistry too. See: Roberts, "Creating Green Chemistry," 38-67.

[69] Previous work on this has also been published: Linthorst, "Origins and Development," 55-68. Section 1.4 outlines the reasons for the focus on these countries.

societies – ACS, RSC and the KNCV – in the emergence and further development of green chemistry. These professional and learned bodies have never been systematically studied and compared in relation to the main topic of this thesis.

Chapters 2, 3 and 4 focus respectively on the USA, the UK, and the Netherlands. Central questions in those chapters are: which (political, governmental, environmental, etc.) factors or actors were most dominant in the institutional development of green chemistry? Next to that I will analyze how different dominant views on what green chemistry is (and should be) emerged and developed. Leading questions for that research objective are: which (political, governmental, environmental, etc.) factors or actors were most dominant in conceptualizing green chemistry? And why and how are these factors or actors dominant? I will show that the answers to these questions have been quite different in the three different national settings.

Chapters 2, 3 and 4 are mainly on how the three national chemical societies and the "green chemistry movement" influences each other, and how different groups and societies tried to "carve out" a place for green chemistry within the disciplinary landscape of chemistry. But these chapters pay less attention to the type of research that was done under the banner of green chemistry. What kind of scientific puzzles did (green) chemists try to solve in the laboratory? What were important theories and guiding principles that played a role in that research? How did those ideas and activities shape the "internal" scientific identity of green chemistry? Authors such as Roberts, Epicoco and Sjöström came to different answers to those questions, in particular with respect to the (disciplinary) role played by the so-called "twelve principles".[70] So, the characterization of green chemistry's internal scientific identity and growth do vary from one STS scholar to another, and range from greenwashing to the presence of a substantial epistemic component. To shed more light on issues such as these I will devote Chapter 5 to a closer look on the twelve principles and the chemical views concerning green chemistry. Next to that Chapter 5 provides an investigation of the internal scientific values and factors that played a role in the shaping and growth of green chemistry.

Theoretical framework

Throughout this thesis I use theoretical concepts from the sociology and philosophy of science for a better understanding of the historical and internal scientific development of green

[70] Compare for instance Sjöström with Epicoco et al.

chemistry. I will shortly summarize the most important ones under three headings: (1) the concept "operational realism", in order to discuss how environmental and green chemistry fit the general character of chemistry; (2) "science and democracy", because Chapters 2, 3 and 4 position the emergence of green chemistry in a broader social, cultural and political framework; and (3) "paradigm choices", because Chapter 5 discusses the epistemic components of green chemistry and the nature of green chemistry as a scientific specialization within the chemical sciences.

A characteristic of chemistry: operational realism

The character of the different natural sciences has often been the subject of philosophical debate. For instance, chemistry has sometimes even been characterized as a kind of magic, because of experimental spectacles for public audiences performed by chemists.[71] Bensaude-Vincent and Jonathan Simon analyzed the public image of the chemical sciences in relation to the identity of chemistry. They argue:[72]

> This image problem is not confined to the chemical industry, however, but also affects the academic discipline of chemistry. The subaltern status of chemistry is as old as the discipline itself. (…) The idea that chemistry is an impure science does not only come from its links to pollution. Chemistry is also considered impure because of its hybrid nature, its constant mix of science and technology.

Based on the 'constant mix of science and technology', Bensaude-Vincent and Simon emphasize the dual nature of chemistry as both a pure and an applied science. They note that there is much philosophical controversy whether chemistry should be ontologically or epistemologically reduced to physics.[73] This endless discussion proved fruitless according to both authors. They postulate a new philosophical approach in order to portray the chemical sciences. That characterization of the chemical sciences is used in this thesis in order to judge to what extent environmental chemistry and green chemistry fit the general picture of the discipline. Chemists, according to Bensaude-Vincent and Simon, are not focusing on the

---

[71] Salim Al-Gailani, "Magic, Science and Masculinity: Marketing Toy Chemistry Sets," *Studies in History and Philosophy of Science* 40 (2009): 372-381.
[72] Bernadette Bensaude-Vincent and Jonathan Simon, *Chemistry: The Impure Science* (London: Imperial College Press, 2008), 2 and 5.
[73] For further reading: Joachim Schummer, "The Philosophy of Chemistry," *Endeavour* 27 (2003): 37-41.

ultimate building blocks that make up our world. Essentially, they accept the existence of the invisible and unobservable entities such as atoms, ions and molecules. This makes chemistry an impure science and that's why it is a kind of science that can 'best [be] qualified as "operational realism" because of its perpetual link to productive activity'.[74] Bensaude-Vincent and Simon define this operational realism as follows:[75]

> Allowing the possibility that elements can evolve along with the relevant techniques, changes their nature. They are no longer conceived as the ultimate building blocks of nature but are instead bound to laboratory operations. This is the first sense of an operational concept; one that does not seek to pick out the essence of the object under consideration, but instead identifies the operations needed to obtain it. (...) What is important for him [Antoine-Laurent Lavoisier (1743-94)] as a chemist is that the elements are actors in chemical operations, and so are defined by how they act and react in a network of relations with other chemical actors.

At the phenomenological level chemists have been analyzing experimental results of their laboratory operations and with their material abstractions as realists they co-produce theory and laws.[76] This knowledge is used by other chemists in the academic arena or from industry in chemical synthesis.

Science and democracy: the communication of contextual values

As is well known, during the Nazi era writing about the theory of relativity and other physical theories (partly) developed by Jewish scientists was forbidden by the regime. Scientists that adhered to the Nazi ideology developed their own 'Deutsche Physik'.[77] This example illustrates that the theory choices made by scientists, and as a result the production of scientific knowledge, can be influenced by ideological and other societal values. Scientists have to find their way not only in the laboratory, but also in society. The relationship between science and society has been analyzed by several STS scholars. Below I discuss three interrelated theories that conceptualize: (1) heuristic agreements between the academic

---

[74] Bensaude-Vincent and Simon, *Impure Science*, 207.
[75] Ibid., 202-203.
[76] For further reading: Alan J. Rocke, *Image & Reality: Kekulé, Kopp, and the Scientific Imagination* (London: The University of Chicago Press, 2010).
[77] Dieter Hoffmann, "Between Autonomy and Accommodation: the German Physical Society During the Third Reich," *Physics in Perspective* 7 (2005): 293-329. See also: Jeroen van Dongen, *Einstein's Unification* (New York, NY: Cambridge University Press, 2010).

enterprise and society (Laurens Hessels and Harro van Lente); (2) distinctive political-cultural influences on policy making (Sheila Jasanoff); and (3) the communication of scientific knowledge within science and with society (Arie Rip and Jan-Peter Voss).

The impacts of contextual influences on the academic enterprise are captured by what Hessels and Van Lente call the 'contract between science and society' (Figure 1.1).[78] I use this concept in my historical analysis, because the previous section (1.2) shows numerous connections between green chemistry and society. Using the "science-society contract" concept of Hessels and Van Lente helps to get a deeper understanding of the role that different views on the rationale and identity of green chemistry might play in creating an interface between green chemistry and society.

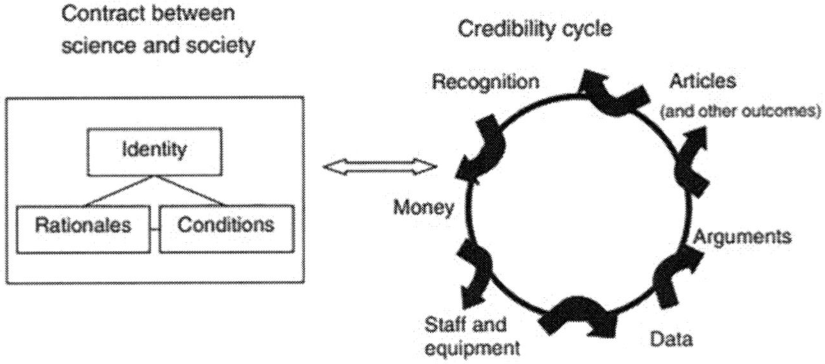

Figure 1.1: The science-society contract (macro level) and credibility cycle (micro level).[79]

Hessels and Van Lente built on the so-called 'credibility cycle' of Bruno Latour and Steve Woolgar, that describes how discoveries made in the laboratory gradually become 'scientific facts' shared by a scientific community.[80] Hessels and Van Lente call these activities the micro level. They are related to and embedded in contracts between science and society, which they call the macro level. That 'macro level' is defined as follows:[81]

---

[78] Laurens K. Hessels and Harro van Lente, "Practical Applications as a Source of Credibility: a Comparison of Three Fields of Dutch Academic Chemistry," *Minerva* 49 (2011): 218.
[79] Ibid.
[80] Bruno Latour and Steve Woolgar, *Laboratory Life: the Construction of Scientific Facts* (Princeton, NJ: Princeton University Press, 1986); Bruno Latour and Steve Woolgar, "The Cycle of Credibility," in *Science in Context: Readings in the Sociology of Science*, ed. Barry Barnes and David Edge (Milton Keynes: The Open University Press, 1982), 35-43.
[81] Hessels and Van Lente, "Practical Applications as a Source of Credibility," 217.

(…) it has three elements, specifying what science should do (identity), why it should do this (rationale), and the appropriate conditions for science to function well (conditions). (…) The contract, that is, the set of implicit and explicit agreements, also describes why science deserves support.

Identity, rationales and conditions, including legitimacy and the types and amounts of funding, for science vary between nations due to differences in societal values and participatory styles of public knowledge and policy making. These kind of influences in democratic political cultures find a place in Jasanoff's concept of civic epistemology:[82]

Civic epistemology refers to the institutionalized practices by which members of a given society test and deploy knowledge claims used as a basis for making collective choices. (…) These collective knowledge-ways constitute a culture's civic epistemology; they are distinctive, systematic, often institutionalized, and articulated through practice rather than in formal rules.

Jasanoff discerns three distinct types of civic epistemologies: 'contentious', 'consensus-seeking' and 'communitarian'.[83] Contentious civic epistemologies are strongly interest-based and follow tacit institutional routines. This type of civic epistemologies heavily relies on the input of interested stakeholders, e.g., industries, activists and academic scientists, in order to produce public knowledge and to make collective choices. This pluralism often comes with distrust and conflicts of interests, which are often legally resolved.[84]

A consensus-seeking civic epistemology, by contrast, is characterized by role-based institutions that negotiate with each other on the basis of mutual trust, and with an active involvement of the state in processes of negotiation.[85] This method for policy making is more broadly conceived than in a contentious civic epistemology, because it is not limited to a few interested stakeholders.

A communitarian civic epistemology also relies on an active involvement of the state, but here public servants of proven expertise and members of trustworthy bodies steer the

---

[82] Sheila Jasanoff, *Designs on Nature: Science and Democracy in Europe and the United States* (Princeton, NJ: Princeton University Press, 2005), 255.
[83] Ibid., 259. Jasanoff compares in her analysis the USA (contentious), the UK (communitarian) and Germany (consensus-seeking). She also 'offers conceptual clarity but at enormous risk of reductionism' and thus warns for a rigid use of her civic epistemology concept.
[84] Ibid., 258-263.
[85] Ibid., 259. Jasanoff characterizes these civic epistemologies also as corporatism.

production of public knowledge and policies. These civil servants sometimes consult the input of the public. Jasanoff characterizes the communitarian political culture as service-based, with top-down policies dominating. Her characterizations of political cultures differ thus with respect to the style of public knowledge-making and the methods of ensuring public credibility, which impact academic enterprises in different ways.

In this thesis I will argue that this applies to the development of green chemistry as well. A historical analysis of the policies concerning science and the environment in three different national settings will help to substantiate that claim. So, in order to characterize socio-political and cultural impacts on the emergence of green chemistry I use the civic epistemology concept of Jasanoff.

Clearly, academic knowledge plays a role in different modern political cultures. In recent decades science policy has been democratized in most of these cultures, and as a result the boundaries between science and society often became somewhat blurred. Considering the "contract between science and society" studied by Hessels and Van Lente, non-specialists are increasingly important members of agencies that provide conditions and rationales for science in modern democracies. Seen in that light, the communication of academic knowledge with society has changed as well, in order to produce the 'implicit and explicit agreements' as referred to in the model of Hessels and Van Lente. Scientists, then, have to show what they are actually "doing" in the laboratory, and with what purposes. This activity of reaching out is interesting with respect to the communicated character of green chemistry, because is it a scientific or social movement, a subfield of chemical science, a meta-discipline, or just a buzzword?

An interesting approach to cope with questions such as these is the concept of 'umbrella term' put forward by Rip and Voss. This concept aims to combine the characteristics of emerging scientific fields with rhetorical devises such as buzzwords that act as mediators. Umbrella terms link scientific promises of new research directions to contextual concerns and science policies.[86] Rip and Voss argue that umbrella terms can be stabilized and reinforced through for instance institutionalization in the organizational field of science (policy). That field can be conceived as a "trading zone" where the linking together of social questions, new scientific knowledge and opportunities takes place.[87] The emergence of umbrella terms can be very specific, as its origins can be based in politics as well as in

[86] Arie Rip and Jan-Peter Voss, "Umbrella Terms as Mediators in the Governance of Emerging Science and Technology," *Science, Technology & Innovation Studies* 9 (2013): 39-59.
[87] Ibid., 41-43.

scientific research, but most importantly, in comparison to mere buzzwords an umbrella term has a substantial epistemic component.[88] According to Rip and Voss, the meaning and 'epistemic status' of umbrella terms are often contested due to, for instance, contextually defined problems.[89] Green chemistry seems to be a very interesting example of an umbrella term in the sense of Rip and Voss, because it combines some characteristics of buzzwords, with a rather robust institutionalization in the form of new institutes and scientific journals, and with an epistemic program directed to the development of more green and sustainable chemical processes.

Academic circles: paradigm choices and "internal" scientific values

Scientific knowledge has the interesting property that it can both lead to concrete technical and practical innovations and to an improved understanding of the world we live in. This knowledge expresses itself in the case of chemistry for instance in theories, scientific laws and, (synthetic) methodologies. It is the daily routine of the scientist to use and test, consciously and unconsciously, implicitly and explicitly, this knowledge in the laboratory. Apart from truth seeking as a scientific virtue, excellent knowledge is characterized by scientific values such as scope, simplicity, accuracy, consistency and explanatory and predictive power.[90] As pointed out by the philosophers of science Martin Curd and Jan Cover, those values might lead to different directions in scientific puzzle-solving or theory choice by scientists.[91] These cognitive phenomena are part of the scientific endeavor and therefore shape the course of science. This section discusses these phenomena with respect to paradigm development, scientific revolutions, and the role of paradigms in scientific specialization.

Thomas Kuhn coined the concept of paradigm, which he defined as 'universally recognized scientific achievements, which for a time provide model problems and solutions to a community of practitioners'.[92] But, the commitment to a certain paradigm, shared by a

---

[88] Bensaude-Vincent, "The Politics of Buzzwords," 240.
[89] Rip and Voss, "Umbrella Terms as Mediators," 53.
[90] In this thesis I use the term internal scientific value, whereas others might prefer the terms epistemic value or cognitive value. Furthermore, try to imagine the scientific power of well-known laws such as those of Newton and Lambert-Beer. See for further reading: Duncan Pritchard, "Recent Work on Epistemic Value," *American Philosophical Quarterly* 44 (2007): 85-110.
[91] Martin Curd and Jan A. Cover, "Commentary," in *Philosophy of Science: the Central Issues*, ed. Martin Curd and Jan A. Cover (New York, NY: W.W. Norton & Company, 1998), 211-253.
[92] Thomas S. Kuhn, preface to *The Structure of Scientific Revolutions*, by Thomas S. Kuhn (London: The University of Chicago Press, 1996), X.

group of scientists, changes due to so-called 'anomalies' that oppose the leading paradigm.[93] This leads to scientific controversy and to a 'new set of commitments' to a newly shared paradigm as opposed to the older one, which is then replaced on the basis of internal scientific values. Kuhn calls this a "scientific revolution".[94]

Ernan McMullin also acknowledges the importance of internal scientific values in a paper on the rationality of paradigm change.[95] He argues that defenders of an old paradigm and offenders of that paradigm, both can have 'good reasons' for supporting one paradigm over the other.[96] But, 'these reasons are never coercive in their own right in forcing change' and should be regarded as values, namely 'internal' scientific ones in this case.[97] By emphasizing rationality McMullin criticizes the incommensurable character of scientific revolutions as proposed by Kuhn. McMullin distinguishes between deep, shallow and aesthetic revolutions. He accepts the option of the introduction of a new paradigm without replacing an older one. This is the so-called "shallow revolution". Just like Kuhn, he accepts the sharing of internal scientific values between competing paradigms. A "deep revolution" takes place when competing paradigms do not share the same set of internal scientific values. But, according to McMullin, deep revolutions hardly occur. When there is no revolution at all, so-called normal science prevails, which Kuhn defines as:[98]

> Normal science consists in the actualization of that promise [of a paradigm], an actualization achieved by extending the knowledge of those facts that the paradigm displays as particularly revealing, by increasing the extent of the match between those facts and the paradigm's predictions, and by further articulation of the paradigm itself.

The progress of normal science can, according to McMullin, be overestimated in scientific debates due to rhetorical and symbolic artefacts.[99] McMullin calls that an "aesthetic revolution".

---

[93] Ibid., 6. For Kuhn, a paradigm can be a law, theory, application and instrumentation altogether.
[94] Ibid. See also: Thomas S. Kuhn, "Objectivity, Value Judgement, and Theory Choice," in *The Essential Tension: Selected Studies in Scientific Tradition and Change*, ed. Thomas S. Kuhn (London: The University of Chicago Press, 1977), 320-339. The importance and role of internal scientific values have been discussed by several scholars. I recommend for further reading: Curd and Cover, "Commentary," 211-253.
[95] Ernan McMullin, "Rationality and Paradigm Change in Science," in *Philosophy of Science*, Curd and Cover, 119-138.
[96] Ibid., 120.
[97] Ibid.
[98] Kuhn, *Scientific Revolutions*, 24-25.
[99] Ibid., 23-42.

As discussed in the previous section (1.2), the characterization of green chemistry differed from one STS scholar to another, with the opposing claims of an aesthetic revolution (greenwashing) and a more profound revolutions implied. The perspective of McMullin is therefore interesting with respect to the revolutionary emergence of green chemistry in order to reflect on its deepness within science. The previous section also shows different characterizations of green chemistry with respect to scientific specialization, often with the role of the "12 principles" included. Therefore I also use the sociological model of Nicholas Mullins in order to estimate to what degree a green chemistry paradigm is shared by social configurations of chemists.

In order to understand the dynamics of science Mullins focuses on the development of scientific specialties, building on Kuhn's ideas.[100] He distinguishes between intellectual and social aspects in the development of science. Intellectual aspects are based on the paradigm the scientists adhere to define scientific puzzles that should be solved. Social dimensions of the development of science are for instance scientific communication, co-authorship, apprenticeship and colleagueship. The social and intellectual aspects are interrelated. Mullins recognizes a pattern of growth that starts with a minimum set of two persons who form a paradigm group. This is followed respectively by a communication network, cluster and, ultimately, a scientific specialty (Figure 1.2). This pattern runs parallel to a growth in scientific consensus among the scientists involved, which is supported by key publications and educational initiatives in order to recruit new members of the scientific specialty.[101] This thesis pays special attention to the intellectual origins of green chemistry in the USA, the UK and the Netherlands and how the ideas and concepts that emerged are mutually related through social processes between the three countries. I will analyze these origins through the lens of Mullins in particular with respect to the first stage of a scientific specialty. He emphasizes the importance of the first stage as follows:[102]

> A very important event for this paradigm occurred when norms were established to govern the kind of research to be done and the manner in which it was to be done and presented.

[100] Nicholas C. Mullins, "The Development of a Scientific Specialty: the Phage Group and the Origins of Molecular Biology," *Minerva* 10 (1972): 51-82.
[101] Arie Rip, "Onderzoek van de Ontwikkeling van Wetenschapsgebieden," *Kennis en Methode* 1 (1977): 318-325.
[102] Mullins, "The Development of a Scientific Specialty," 56.

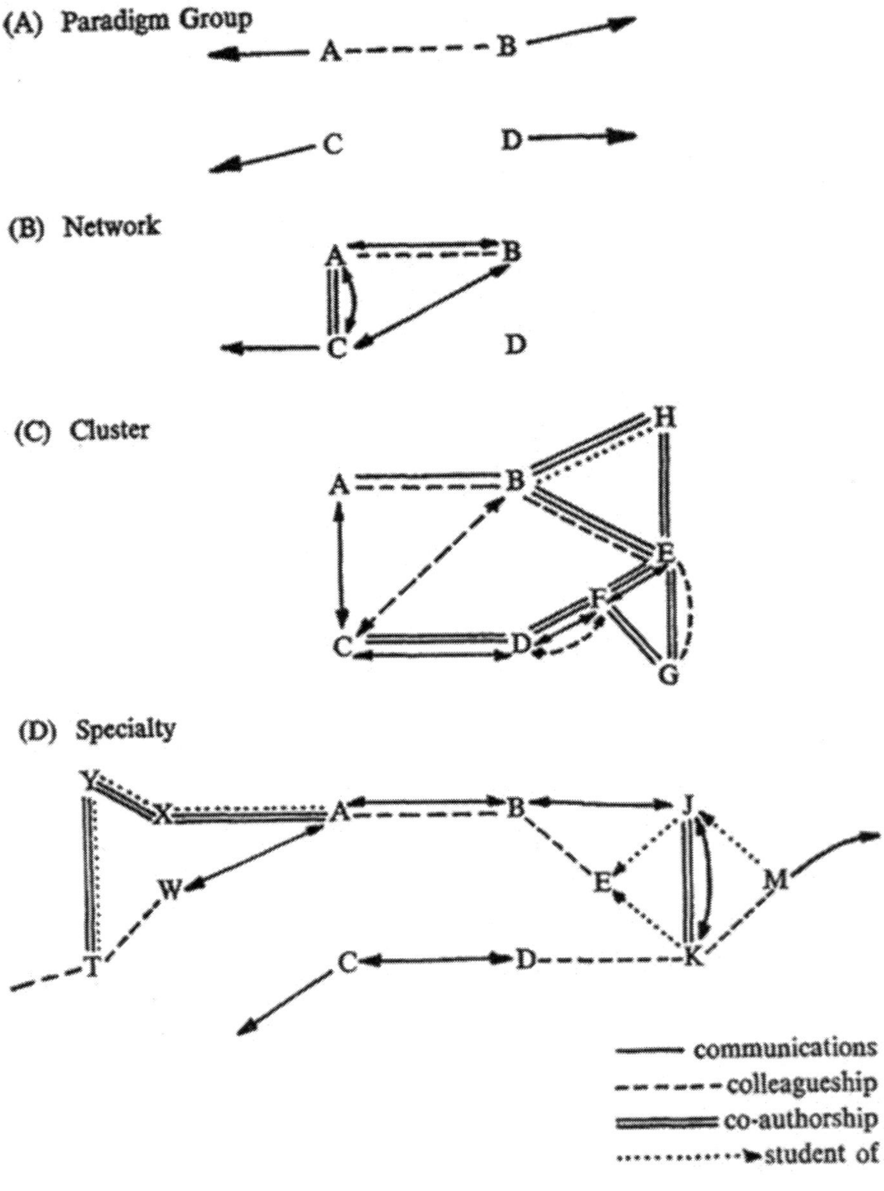

Figure 1.2: The development of a scientific specialty.[103]

Finally, a scientific specialty is formed that has evidently more social and 'formal organization' than previous stages, as witnessed by the establishment of specific institutions, laboratories and academic journals.[104] In a scientific specialty, Mullins states, scientists 'may

[103] Ibid., 53.
[104] Ibid., 74.

share a paradigm and a set of judgements about what general work should be done in the field, although the details of those ideas might differ'.[105] Scientists are then intellectually connected. The whole development towards a new specialty takes about 25 years, according to Mullins.[106]

1.4 Methodology

Jody Roberts argued that dominant green chemists from the USA have been strategically writing their own historical narrative of environmental problems, the public image of chemistry and the chemical sciences. The outcome of that story was a plea for "green chemistry". That strategy reinforced the frame used by those actors in order to persuade other chemists to join the green chemistry community.[107] I consider that historical development to be a warning for my historical analysis of the development of green chemistry, as primary sources written by dominant actors might be colored. The methodology of the research presented in this book therefore relies on different research approaches and a wide range of sources, not solely those from dominant actors.[108] I use sources that are directly related to the ACS, RSC and KNCV, which provide the basis for my comparative analysis. Next to that I use scientific articles concerning green chemistry as input for quantitative and qualitive research methods, which I will explain below.

I argued in section 1.2 that the majority of STS scholars focused on the USA, but in this thesis the origins of green chemistry will be investigated on the basis of a comparative analysis of three countries. Because, following Jasanoff, the production of academic knowledge is shaped differently from one civic epistemology to another: for instance, being contentious (USA), or communitarian (UK). She stresses the importance of comparative analyses in order to examine the development of knowledge societies.[109] I follow that example, and compare the three countries from the 1960s onwards with respect to changing conditions and rationales for the chemical sciences (see also Hessels and Van Lente). I have chosen these countries, because in an earlier publication I have shown that green chemistry

---

[105] Ibid. The findings of Mullins are comparable to those of Frickel and Gross.
[106] Nowadays, I assume 25 years is a bit too long as the rate of scientific communication has been increased since the seminal work of Mullins due to the world wide web and social media.
[107] Roberts, "Creating Green Chemistry," 38-67.
[108] Rip emphasizes the importance of studying science as an object from perspectives, ranging from social and political history to internalist oriented developments of science (internalism): Rip, "Onderzoek van de Ontwikkeling van Wetenschapsgebieden," 311-333.
[109] Jasanoff, *Designs on Nature*. See also the previous section.

started in the mid-1990s in the USA and has been prominent since then.[110] Major efforts of green chemistry labeled activities have also been observed for the UK though, which has a rich tradition in the chemical sciences too.[111] The USA and the UK will therefore extensively be analyzed and compared in this thesis, and the Netherlands will be taken on board as an example of a Continental European country (cf., Jasanoff).[112]

Government policies concerning the environment and science have influenced the development of the discipline. In the next chapters I will show that these policies also had an influence on decisions taken by the chemical societies. These learned bodies were important actors that facilitated and influenced dialogues among chemists at for instance national conferences and symposia. Chemical societies united members from government, industry, university and education. In all three countries these societies are organized in divisions, committees and working parties that all have special missions, or address different subfields of chemistry. As mentioned above, in this thesis, the focus is on the national chemical societies ACS, RSC and KNCV, and their divisions or working parties in fields of environmental chemistry and green chemistry. These professional and learned bodies are permanently in dialogue with society, policymakers and politicians about government policies and changing contextual values and the role of chemistry herein. In this thesis I will compare those "dialogues" for the USA, the UK and the Netherlands in order to determine similarities and differences related to the emergence of green chemistry.

My comparison is based on annual reports of the ACS, RSC, CS and KNCV, as well as news magazines of these societies.[113] The majority of the older reports were consulted as part of archival research at the headquarters of the ACS (Washington, DC), the RSC and CS (both London) and the KNCV (Leidschendam). Archival research has also been carried out at the Science History Institute (Philadelphia, PA), which houses a wide range of minutes and newsletters of DEC from the 1980s onwards that had not been subject to historical investigation. Similar primary sources were analyzed during my archival research in London

[110] Linthorst, "Origins and Development," 55-68.
[111] Ibid.
[112] Owing to for example the Dutch Nobel prize winners Jacobus Henricus van 't Hoff (1852-1911), Petrus Josephus Wilhelmus Debye (1884-1966), Paul J. Crutzen (1933-2021) and Bernard L. Feringa (1951-), the small country of the Netherlands has a rich and international tradition in the chemical sciences.
[113] For ACS: www.acs.org; for RSC: www.rsc.org; and, for KNCV: www.kncv.nl (all accessed May 22, 2022).

and Leidschendam.[114] Additionally, interviews were conducted with important and dominant actors of the EPA, ACS, RSC and KNCV.[115]

Next to that, quantitative methods and scientometric analyses of bibliographic data of green chemistry publications were carried out by using search engines such as ISI Web of Knowledge: Web of Science. One of my studies concerns the scientific articles published in the first volume of *Green Chemistry* in order to determine the bibliographic coupling of these articles.[116] Following Henry Small, scientific articles can be 'conceived as a dialogue among citing authors on the "meaning" of earlier texts' in the construction of knowledge.[117] The point of departure here is that authors who are citing the same document(s) are concerned with similar scientific puzzles, and probably join the same scientific specialty.[118] The analysis of bibliographic coupling thus provides new insights on to what extent green chemists are intellectually associated, in particular at the start of green chemistry publishing. Next to that the number of publications on green chemistry per year gives a proxy for the growth of the field. Both these analyses are of additional help in understanding the emergence of green chemistry, next to qualitative methods.

Discourse analysis was done with a special focus on key actors and their publications on green chemistry in academic journals. I also pay attention to differences in scientific meaning and to various (chemical) goals and views by using conceptual analyses.[119]
All research outcomes are finally placed and characterized in their historical context, by combining internal and contextual developments of science.

1.5 Structure of this Thesis

The next chapters (Chapters 2, 3 and 4) of this book discuss institutional developments regarding green chemistry against the background of environmental problems, government policies and the public image of chemistry in respectively the USA, the UK and the Netherlands. Each chapter is structured in a similar way and discusses contextual influences

[114] See "Archives and Collections" in the chapter Sources and Bibliography.
[115] See "Interviews and Personal Communications" in the chapter Sources and Bibliography.
[116] Network visualization took place with Pajek. See heretofore: "Hands-on Class on Science & Technology Indicators," Loet Leydesdorff, accessed May 22, 2022, http://www.leydesdorff.net/indicators/lesson1.htm.
[117] Henry G. Small, "Cited Documents as Concept Symbols," *Social Studies of Science* 8 (1978): 327.
[118] See also: Henry G. Small, "A Co-citation Model of a Scientific Specialty: A Longitudinal Study of Collagen Research," *Social Studies of Science* 7 (1977): 139-166.
[119] Sjaak Koenis and Karin Bijsterveldt, *Conceptual Analysis: an Introduction*, unpublished. This document is used within the Arts & Sciences Program of Maastricht University (The Netherlands).

on the emergence of green chemistry: (1) the political and cultural attention for the environment; (2) the rise of environmental concerns within chemical societies, demonstrated especially by the creation of environmental chemistry divisions; (3) the rise of pollution prevention policies; and (4) institutional developments promoting green chemistry against the background of culturally conditioned government polies.

Dominant green chemistry developments started in the USA (Chapter 2). This chapter emphasizes the role of the EPA with respect to the institutionalization of green chemistry, in the context of the political culture of the USA concerning the environment. Chapter 3 discusses the UK. It is in this country that the first major internationally oriented scientific journal on green chemistry was founded, published by the RSC. Interestingly, with respect to the launching and editorial work of this journal, a research group of one particular UK university (York) proved influential. Dutch chemists were part and parcel of the international science of environmental chemistry since its existence. This is analyzed in Chapter 4. It shows that in the Netherlands the term "sustainable chemistry" became popular. The rise of this term is analyzed in the context of the government policy programs of the Dutch government. The (chemical) views, ideas and concepts concerning green chemistry proposed by the dominant actors are discussed in Chapter 5. Finally, the research findings of Chapters 2 to 5 are combined and discussed in Chapter 6 in order to draw conclusions with respect to the main research questions.

Chapter 2: Chemistry, Politics and Society: the Rise of Green Chemistry in the USA

In this chapter I will put forward the argument that the "birth" of green chemistry took place in the USA. In this and the next two chapters I will analyze the development of environmental chemistry and green chemistry in each country through the lens of national chemical societies. This chapter focusses on the American Chemical Society (ACS).

With more than 150,000 members, the ACS is the world's largest scientific society in the chemical arena.[120] For many decades, this learned and professional body has been representing chemists working in industry, governmental agencies, education and universities. The ACS has been publishing journals and books on a wide range of chemical topics, and it has a large international footprint in publishing on chemistry. That fact is important for understanding the international emergence of green chemistry.

The way the ACS coped with environmental pollution and with the rise of green chemistry against the background of social concerns is highly relevant for the topic of this book. In this chapter, I will focus on the responses of the ACS to environmental concerns and on the way the society tried to influence the public image of chemistry from the 1960s onwards. The leading question will be which factors influenced the growth and recognition of green chemistry within the ACS. I will argue that important developments concerning the green chemistry movement in the USA must be understood against the background of a growing public environmental concerns, together with the corresponding political responses. The ACS created at an early date a Division that focused on environmental chemistry. Within that division, a green chemistry movement developed and this was recognized at the highest policy levels of the ACS.

2.1 The Emergence of Environmental Policies during the 1960s-1980s

Since over a century the USA had to cope with environmental problems.[121] It was felt that the growing population and its economic consequences negatively influenced the environment. In 1962, the famous biologist Rachel Carson greatly fueled this concern with her book *Silent*

---

[120] "About ACS," ACS, accessed May 22, 2022, https://www.acs.org/content/acs/en/about.html.
[121] Carolyn Merchant, ed., *Major Problems in American Environmental History* (Lexington, MA: D.C. Heath and Company, 1993).

Spring.[122] She warned for the widespread use of pesticides and its consequences to nature and mankind. Other persons and groups were concerned about the pollution of air, water and soil due to industrial activities. Large oil spills showed the public the devastation of nature. In fact, based on fossil feedstocks, USA's production of synthetic organic chemicals increased considerably from the 1950s onwards.[123] Some of these manmade chemicals had not been observed in nature before. As a result, in the 1960s environmental quality had become a serious problem, also in the view of large parts of the population.

This coincided with social changes, when people started struggling for civil rights.[124] The 1960s were also a period in which USA's economy boomed. Unemployment was low, as was the Federal Debt (Figure 2.1). The greater economic welfare led to increasing 'concern[s] over the quality of life'.[125] And the culture of the civil rights movement encouraged people 'to take direct action to solve society's ills'.[126] Scientific knowledge on the environment also grew in the same period. As a result, educated people increasingly organized themselves as environmental interest groups.[127] In the field of occupational health and safety now also trade unions became increasingly involved, leading for instance to public hearings on air quality standards.[128] The growing concern on the environment culminated in 1970 when some 20 million participants took part in celebrating the first Earth Day.[129]

---

[122] Rachel Carson, *Silent Spring* (New York, NY: Houghton Mifflin Company, 1962). See also: David K. Hecht, "Constructing a Scientist: Expert Authority and Public Images of Rachel Carson," *Historical Studies in the Natural Sciences* 41 (2011): 277-302.
[123] Donald G. Crosby, "Review: Environmental Chemistry: an Overview," *Environmental Toxicology and Chemistry* 1 (1982): 1-8.
[124] Mary B. Norton et al., *A People and a Nation: a History of the United States* (Boston, MA: Houghton Mifflin Company, 1991), 545-569; Morton Keller, *America's Three Regimes: a New Political History* (New York, NY: Oxford University Press, 2007).
[125] Riley E. Dunlap and Angela G. Mertig, "The Evolution of the U.S. Environmental Movement from 1970 to 1990: an Overview," in *American Environmentalism: the U.S. Environmental Movement, 1970-1990*, ed. Riley E. Dunlap and Angela G. Mertig (Washington, DC: Taylor & Francis New York, 1992), 3.
[126] Ibid., 2.
[127] Robert C. Mitchell, Angela G. Mertig and Riley E. Dunlap, "Twenty Years of Environmental Mobilization: Trends Among National Environmental Organizations," in *American Environmentalism*, Dunlap and Mertig, 11-26.
[128] Anonymous, "Steelworkers to Act on Air Pollution," *Chemical & Engineering News* 47 (1969): 22; Daniel Faber and James O'Connor, "Environmental Politics," in *Major Problems*, Merchant, 551-559.
[129] Faber and O'Connor, "Environmental," 551-559; Dunlap and Mertig, "The Evolution of the U.S. Environmental Movement," 2.

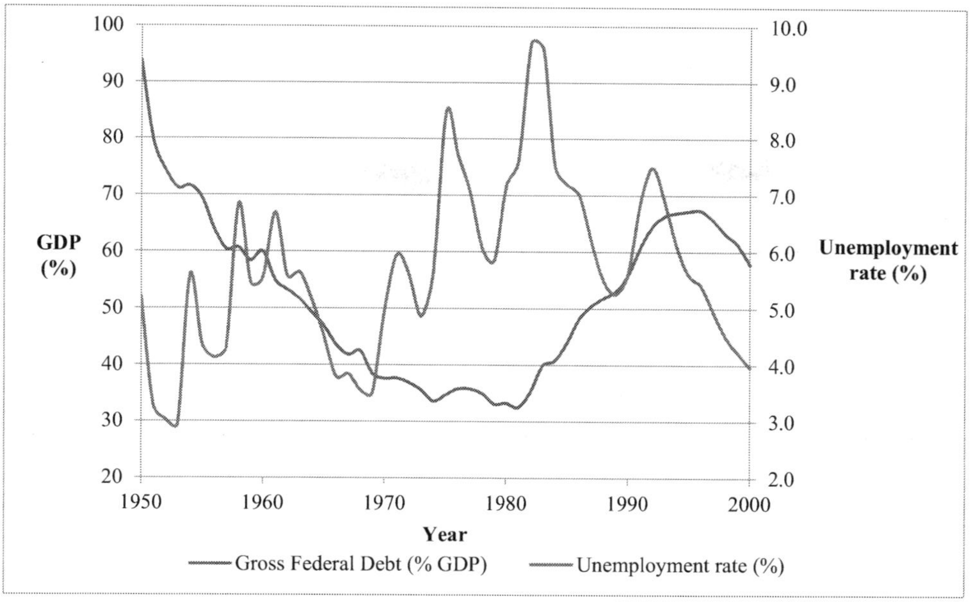

Figure 2.1: Gross Federal Debt (% GDP) and unemployment rate (%) for the 1950-2000 period.[130]

The environmental concerns among the population gave impulses to the enactment of new environmental laws.[131] In 1969 US Congress passed the National Environmental Policy Act (NEPA) under the Republican Administration of President Richard Nixon.[132] This federal act politically confirmed the mutual relationship between public health, nature conservation and environmental protection. It marks a major shift in environmental policy. The conservation of wildlife areas and the environment in general was not sufficient anymore. Protecting public health now became an important concern. The Nixon Administration recognized that environmental governance was handled by many unconnected departments, agencies and offices. It therefore came to the conclusion that more coordination was necessary. Consequently, the Council on Environmental Quality was installed, as a result of

[130] % GDP data are electronically obtained from the US Census Bureau History and unemployment rate data from the US Department of Labor, section Bureau of Labor Statistics.
[131] Samuel P. Hays, "From Conservation to Environment: Environmental Politics in the United States since World War II," *Environmental Review* 6 (1982): 14-41; Michael S. Dukakis, "Environmental Politics in Post-World War II America," *Resources, Conservation and Recycling* 18 (1996): 5-9. In 1988, Dukakis was the Democrat's Candidate for President of the USA.
[132] "The National Environmental Policy Act of 1969," in *Major Problems*, Merchant, 498-500.

NEPA, within the Executive Office of the President.[133] Moreover, the Nixon Administration unified the majority of federal pollution-control activities in a new independent agency: the US Environmental Protection Agency (EPA), founded in 1970, with William D. Ruckelshaus as its first Administrator.[134] As we will see below, the EPA became very important in the field of environmental control and research.

Ruckelshaus later summarized: '[US] Congress passed a cascade of laws to regulate air pollution, water pollution, solid waste disposal, radiation, noise, pesticides and toxic substances', within a relatively short time.[135] For example, the Clean Air Act of 1970 (CAA) resulted in national ambient air quality standards for hazardous compounds.[136] In 1976, two more acts passed US Congress on the federal control and regulation of hazardous substances: the Solid Waste Disposal Act and the Toxic Substances Control Act (TSCA), of which the first is better known as the Resource Conservation and Recovery Act (RCRA). TSCA regulated the pollution of the environment and the exposure of the public to toxic chemicals that were not covered by other environmental laws. The TSCA enabled EPA to control manufacturers of (toxic) chemicals in a far-reaching manner, for instance by banning of manufacture of chemicals that posed serious threats to the environment and public health.[137] Triggered by the Love Canal disaster of the late 1970s, environmental legislation in 1980 got a new dimension with the so-called 'Superfund', which regulated issues concerning (private) compensation and liability.[138] Now also 'individuals' were given the legal opportunities to

---

[133] "Council on Environmental Quality," The Administration, accessed May 22, 2022, https://www.whitehouse.gov/ceq/.

[134] Hays, "From Conservation to Environment," 14-41; Dukakis, "Environmental Politics," 5-9; Robert Gillette, "Environmental Protection Agency: Chaos or Creative Tension," *Science* 173 (1971): 703-707. See also "EPA History," EPA, accessed May 22, 2022, https://www.epa.gov/aboutepa/epa-history.

[135] William D. Ruckelshaus, "Environmental Protection: a Brief History of the Environmental Movement in America and the Implications Abroad," *Environmental Law* 15 (1985): 455-469. Ruckelshaus left EPA in 1973, but was reappointed in 1983 by the Reagan's Administration.

[136] Peter H. Spitz, "The Chemical Industry and the Environment: Meeting the Challenge," in *The Chemical Industry at the Millennium: Maturity, Restructuring, and Globalization*, ed. Peter H. Spitz (Philadelphia, PA: Chemical Heritage Press, 2003), 206-246.

[137] For a detailed description of the CAA, the Solid Waste Disposal Act and the TSCA, see: Committee on Commerce U.S. House of Representatives, *Compilation of Selected Acts within the Jurisdiction of the Committee on Commerce: Environmental Law* (Washington, DC: US Government Printing Office, 1995).

[138] Superfund is the common name for the Comprehensive Environmental Response, Compensation, and Liability Act of 1980. See also: Andrew J. Hoffman, *From Heresy to Dogma: an Institutional History of Corporate Environmentalism* (San Francisco, CA: The New Lexington Press, 1997), 166-167.

hold a company financially responsible when their property was affected by a company's waste.[139]

The environmental quality standards, the launching of new far-reaching environmental acts and the establishment of EPA characterize USA's environmental policy style of the 1970s: a policy of command-and-control.[140] All this led to an increase of EPA's regulating actions from 1971 onwards. The so-called administrative enforcement's actions, for instance, grew from zero, in 1971, to 2,750 per year, in 1976. Next to that in 1980 also approximately 225 civil enforcement actions took place.[141] EPA thus proved to be an active agency as it developed as one of the 'weapons of the environmental movement'.[142] But companies often fought back, and the number of lawsuits between the EPA and chemical industries grew significantly during the 1970s.[143] As a result a culture of polarization and distrust between government, industries and the public evolved in this domain.[144]

Around 1980 though, the intensity of the polarization on environmental issues decreased, mainly due to political and economic changes. The years of Reaganomics had begun. Riley Dunlap and Angela Mertig, two environmental sociologists, reviewed several polls held among USA citizens. They found that environmental concerns moderately decreased in the course of the 1970s, although they remained substantial.[145] At the same time the economy deteriorated, and unemployment rates grew dramatically (cf., Figure 2.1). As a result the public concerns about the economy grew. In 1981, Ronald Reagan (Republican) politically benefitted from that situation and became the new President of the USA.[146] On the basis of his conservative agenda, he wanted to improve market forces and supply-side economics. Consequently, government spending to social programs decreased, federal

---

139 "Comprehensive Environmental Response, Compensation, and Liability Act of 1980," in *Compilation of Selected Acts*, Committee on Commerce U.S. House of Representatives, 861.
140 This policy is also called end-of-pipeline. Joel A. Tarr, "Industrial Waste Disposal in the United States as a Historical Problem," *Ambix* 49 (2002): 4-20. See also: David G. Stephan and John Atcheson, "The EPA's Approach to Pollution Prevention," *Chemical Engineering Progress* 85 (1989): 53-58.
141 Hoffman, *From Heresy to Dogma*, 113.
142 Faber and O'Connor, "Environmental," 555.
143 Hoffman, *From Heresy to Dogma*, 89.
144 Ibid., 64-65; Sheila Jasanoff, *Designs on Nature: Science and Democracy in Europe and the United States* (Princeton, NJ: Princeton University Press, 2005), 247-271; Spitz, "The Chemical Industry and the Environment," 218. Norton et al., reported a huge general distrust of the American public in the USA government that was highlighted by the Watergate scandal: Norton et al., *A People and a Nation*, 570-574.
145 Dunlap and Mertig, "The Evolution of the U.S. Environmental Movement," 1-10.
146 Ibid. See also: Norton et al., *A People and a Nation*, 554-607.

research budgets came under pressure and taxes for the wealthy and corporations were lowered.[147]

Environmental policies also changed because environmental legislation and control had financial consequences for chemical and other industries in order to comply with environmental quality standards. The chemical industry was therefore opposing to strict environmental legislation. Reagan responded with a huge cut in EPA's budget. As a result, EPA enforcement actions dropped immediately and so did the subsidizing of environmental research by EPA (cf., Figure 2.2).[148] The 'conditions' for scientific research on chemistry and the environment deteriorated.[149]

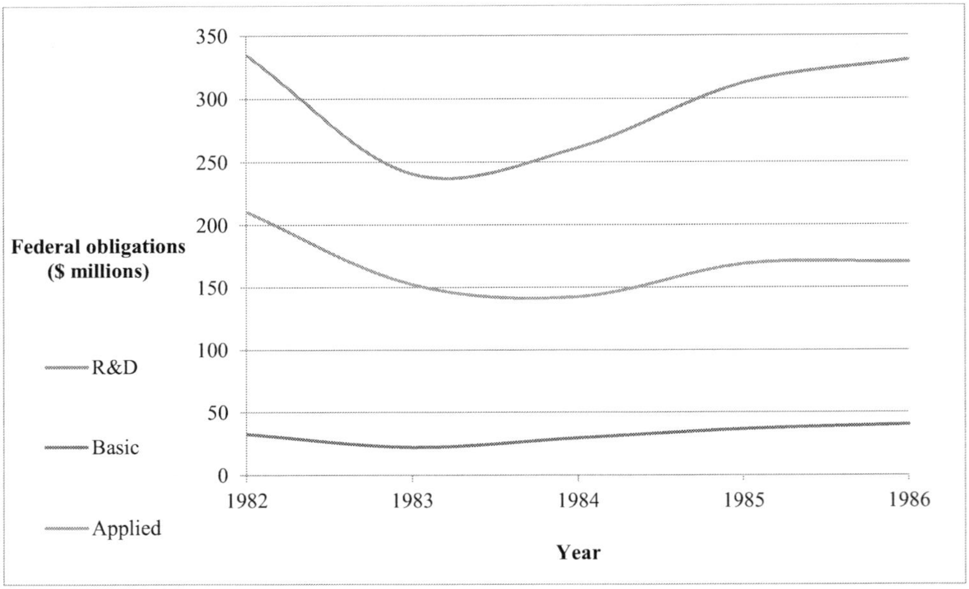

Figure 2.2: Federal obligations for R&D, basic research and applied research regarding the environment and public health (EPA).[150]

[147] For an overview of federal R&D budgets, see: Gregory J. Hather et al., "The United States of America and Scientific Research," *Plos One* 5 (2010): 1-9. See also: Richard Arnold and Andrew B. Whitford, "Organisational Dilemmas of the US EPA: Why Structures Matter for Environmental Protection," *Environmental Politics* 14 (2005): 118-123.

[148] Hoffman, *From Heresy to Dogma*, 113.

[149] Laurens K. Hessels and Harro van Lente, "Practical Applications as a Source of Credibility: a Comparison of Three Fields of Dutch Academic Chemistry," *Minerva* 49 (2011): 218. See Chapter 1 for the "science-society contract".

[150] Data are obtained from: Anonymous, "Facts & Figures for Chemical R&D," *Chemical & Engineering News* 64 (1986): 32-60. Here, the 1986 values were estimated by the National Science Foundation (NSF).

In the wake of Reagan's new policies, the polarization between the EPA and the chemical industry decreased. Anne Gorsuch, the new EPA Administrator appointed by Reagan, got the task to implement a policy that would be more friendly to the (chemical) industry. She 'immediately announced that she wanted cooperation rather than confrontation with the regulated community'.[151] Not surprisingly, environmentalists heavily criticized her policy style for being too close to the industry that had to be regulated, and also for a lack of transparency. Under the pressure of the media Gorsuch resigned her post in 1983.[152] For electoral reasons Reagan thereupon immediately increased EPA's budgets (cf., Figure 2.2) and improved public transparency with the Superfund Amendments and Reauthorization Act of 1986.[153]

In the meantime, there were clear indications that the quality of the environment was improving. For instance, in 1990, SO_2 emissions had dropped by 21.7 % as compared to 1980.[154] Despite these environmental improvements administrative enforcement actions by EPA grew to approximately 3,800 per year in 1990. Civil enforcement actions peaked highest in 1990, and EPA increasingly employed enforcement actions using criminal law. Federal lawsuits with environmental groups as plaintiff grew too.[155] These environmental groups grew in size, and had millions of US citizens behind them.[156] As a result, around 1990, environmental problems were seen as important social concerns and were often the subject of lawsuits. After a period in which polarization between industry, government and society decreased, by 1990 it was apparently growing again. This set the stage for developments within the chemical science during the 1990s that will further be discussed below.

---

[151] Hoffman, *From Heresy to Dogma*, 87; Theodore Goldfarb, "Environmental Legislation," in *Major Problems*, Merchant, 547-551.
[152] See for example: Anonymous, "Washington: Bad Choice, Worse Timing," *Time*, July 16, 1984, accessed May 22, 2022, http://content.time.com/time/magazine/article/0,9171,951225,00.html.
[153] "Superfund Amendments and Reauthorization Act of 1986," in *Compilation of Selected Acts*, Committee on Commerce U.S. House of Representatives, 985-1039. This act is also known as the Emergency Planning and Community Right to Know Act.
[154] "Sulfur Dioxide Trends," EPA, accessed May 22, 2022, https://www.epa.gov/air-trends/sulfur-dioxide-trends.
[155] Hoffman, *From Heresy to Dogma*, 110-113.
[156] Dunlap and Mertig, "The Evolution of the U.S. Environmental Movement," 1-10; Russell J. Dalton, "The Greening of the Globe? Cross-national Levels of Environmental Group Membership," *Environmental Politics* 14 (2005): 441-459; Riley E. Dunlap, "Public Opinion in the 1980s: Clear Consensus, Ambiguous Commitment," *Environment* 33 (1991): 10-15 and 32-37.

2.2 The Emergence of Environmental Chemistry within ACS and the Influence of EPA

The growing attention for environmental issues also took place inside the ACS. In 1970 the ACS Division of Pesticide Chemistry was approved full status as a technical division; in 1973 the ACS Division of Water, Air, and Waste Chemistry was renamed the ACS Division of Environmental Chemistry (DEC); in 1980 the ACS Division of Chemical Health and Safety received full status as a technical division; and in 1981 the same happened with the ACS Division of Geochemistry. The roots of the largest environmentally related division, on environmental chemistry, did even go back to 1915 when the ACS Division of Water, Sewer, and Sanitation Chemistry was created. Since 1964 it had been named the ACS Division of Water, Air, and Waste Chemistry.[157] In the present section I analyze the growth of DEC from about 1970 to 2000.

In response to the debates on a wide range of environmental concerns of the 1960s the ACS published in 1969 *Cleaning our Environment: the Chemical Basis for Action*, with the aim to reposition chemistry from being a cause of environmental problems to being part of the solution.[158] It was reprinted several times and, according to the President of the ACS, used by at least 134 American universities and colleges for educational purposes. More than 45,000 copies were printed, and part of those were send to US Congress and to governmental agencies.[159] The report marks the fact that the ACS, as a professional and learned body, felt challenged by social and political debates on the environment. *Cleaning our Environment* was not primarily focused on environmental experts in the academic arena, but also on 'the involved and educated layman', as was argued by Franklin Long, member of the ACS Subcommittee on Environmental Improvement.[160]

This subcommittee, with members from both industry and academia, had a leading role in writing *Cleaning our Environment*. The so-called Task Force on Environmental Improvement, with members from the ACS Division of Water, Air, and Waste Chemistry, provided this committee with scientific knowledge. Dozens of scientists were recruited from

---

[157] Harold E. Pattee, "ACS Divisions," in *The American Chemical Society at 125: a Recent History, 1975-2001*, ed. Kenneth M. Reese (Washington, DC: ACS, 2002), 78; DEC, "Division of Water, Air, and Waste Chemistry," April 10, 1973.

[158] ACS Subcommittee on Environmental Improvement and ACS Committee on Chemistry and Public Affairs, *Cleaning our Environment: the Chemical Basis for Action* (Washington, DC: ACS, 1969). This report concerns pesticides, solid wastes and environmental quality of air and water.

[159] Wallace R. Brode, "President's Message" in "American Chemical Society: 1969 Annual Report," *Chemical & Engineering News* 48 (1970): 2a-3a; ACS, "American Chemical Society: 1972 Annual Report," *Chemical & Engineering News* 51 (1973): 27.

[160] Franklin A. Long, foreword to *Cleaning our Environment*, V.

chemical industries, universities and federal agencies for this taskforce. These scientists covered a variety of scientific fields, such as toxicology, ecology and analytical chemistry. That interdisciplinary background supported the goal of *Cleaning our Environment* to 'set down an objective account of the current status of the science and technology of environmental improvement'.[161] Seventy-three recommendations were formulated with respect to environmental control. The general view disseminated was to base governmental action on sound science and technology, embedded in federal law. To this end, the report called for more applied and basic research: basic, because the environment in all its complexity is difficult to understand; applied, because new knowledge and technologies were necessary that could be used in existing administrative and legal measures.

Available chemical knowledge allowed the ACS to announce that 'the promulgation of air quality criteria by the Federal Government should be supported in a manner that will allow it to proceed with all possible speed'.[162] During those years, as described above, federal actions followed and end-of-pipe environmental legislations were introduced. These measures went hand in hand with a demand for chemical knowledge concerning the environment. Figure 2.3 shows that also the interest of chemists in the environment increased throughout the 1970s and 1980s, as witnessed by increasing membership numbers of DEC as well as its share in the total membership of the ACS.

A frequently cited article by the former EPA officer Larry Keith (Chairman of DEC) and William Telliard (EPA) illustrates that pattern.[163] In 1979, they reviewed several analytical practices, e.g., gas chromatography-mass spectrometry and inductively-coupled plasma argon emission spectrometry, which were employed by environmental chemists in the analysis of a wide range of toxic pollutants in industrial wastewater discharges. Why? In 1978 EPA was forced to court by environmentally concerned plaintiffs whom argued that EPA failed in the implementation of the Federal Water Pollution Control Act of 1948 due to the 1972 amendments made to this law, which implied the imposition and maintaining of new

[161] Lloyd M. Cooke, preface to *Cleaning our Environment*, VII. Cooke (Union Carbide Corp.) was Chairman of the ACS Subcommittee on Environmental Improvement.

[162] ACS Subcommittee on Environmental Improvement and ACS Committee on Chemistry and Public Affairs, *Cleaning our Environment*, 9. Other environmental topics concerned for instance the banning and reduction of lead in gasoline.

[163] Larry H. Keith and William A. Telliard, "ES&T Special Report: Priority Pollutants I. A Perspective View," *Environmental Science and Technology* 13 (1979): 416-423. According to Scopus, this article has been cited more than 1,950 times (Accessed May 22, 2022).

strict environmental quality standards.[164] The complaints of the environmental interest groups about the amendments were adopted by the court and led to the so-called EPA Consent Decree, which urged EPA to focus on a wide range of individual chemicals in wastewater instead of general environmental quality parameters. In order to fulfil that need, (applied) research on measuring techniques was required, based on knowledge of environmental chemistry. Keith and Telliard's article also marks the impact of EPA within DEC: EPA officers were part of this scientific network and played prominent roles.[165]

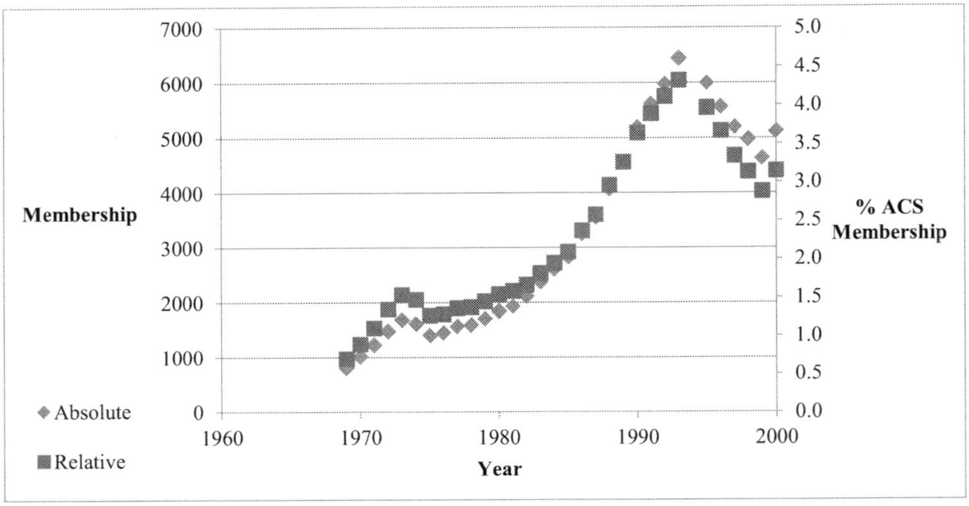

Figure 2.3: Environmental chemistry membership (DEC) within the ACS.[166]

Because of the importance of chemical techniques in end-of-pipe analyses of hazardous substances, DEC-symposia were often organized in cooperation with the ACS Division of Analytical Chemistry. Some of these meetings also concerned basic research questions on, for instance, the pathways of chemicals in the environment under the influence of geological, meteorological and other natural processes. In 1974 a symposium was therefore

[164] In later years the law became commonly known as the Clean Water Act. See: Hoffman, *From Heresy to Dogma*, 64-67; "History of the Clean Water Act," EPA, accessed May 22, 2022, https://www.epa.gov/laws-regulations/history-clean-water-act.

[165] For another article with strong EPA involvement: Lawrence H. Keith et al., "Principles of Environmental Analysis," *Analytical Chemistry* 55 (1983): 2210-2218. Most likely, Lawrence and Larry are the same persons.

[166] The data were received from the ACS Office of Public Affairs.

organized in cooperation with the ACS Division of Colloid and Surface Chemistry and, in 1978, with the Society of Environmental Geochemistry and Health.[167]

It is no surprise therefore that in 1982, the environmental chemist Donald Crosby concluded at the occasion of the launch of the new journal *Environmental Toxicology and Chemistry*, that environmental chemistry had a heterogeneous identity:[168]

> Of course, environmental chemistry includes the manmade pollutants, but risk analysis of such chemicals cannot be separated from the context of Nature. It involves a dynamic global system which includes chemicals in food as well as in waste dumps, toxic plants as well as factory effluents, mussel poisons and the Mussel watch.

By the late 1980s DEC defined environmental chemistry as a 'chemistry which occurs in a variety of media; e.g., air, water, soil, and plant and animal tissue, and includes a broad range of chemistry and related sciences within its scope'.[169] Environmental chemists were active in a diversity of environmentally related research areas, e.g., analytical chemistry, geochemistry, toxicology, engineering, environmental legislation and regulation, and chemicals and the environment. In contrast with the general "operational realism" character of chemical science put forward by Bernadette Bensaude-Vincent and Jonathan Simon though, environmental chemists were not aiming for the design of new (industrial) chemicals and syntheses in their production of chemical knowledge.[170]

Since 1990, divisional symposia at ACS National Meetings have been based on an overall thematic programming. That gave a new dimension to the style in which DEC operated. In August 1990, at the Washington meeting, the first symposium following that

---

167 Victor R. Deitz, ed., *Removal of Trace Contaminants from the Air* (Washington, DC: ACS, 1975); Everett A. Jenne, ed., *Chemical Modeling in Aqueous Systems: Speciation, Sorption, Solubility, and Kinetics* (Washington, DC: ACS, 1979). See also for example: Rod G. Zika and William J. Cooper, ed., *Photochemistry of Environmental Aquatic Systems* (Washington, DC: ACS, 1987); Reese, *The American Chemical Society*, 76-149. See Figure 2.2 for the impact of basic research within EPA.
168 Crosby, "Review: Environmental Chemistry," 7-8. See for similar views: John W. Moore and Elizabeth A. Moore, "Resources in Environmental Chemistry: I. an Annotated Bibliography of the Chemistry of Pollution and Resources," *Journal of Chemical Education* 53 (1976): 167-171; William H. Glaze, "Positioning Environmental Chemistry," *Environmental Science & Technology* 23 (1989): 1173.
169 DEC, *ACS Division of Environmental Chemistry Secretary's Report for the Year 1988* (Washington, DC: ACS, 1989), 1. This definition is consistent with the one of Manahan in his well-known, and perhaps also the oldest, handbook on environmental chemistry: Stanley E. Manahan, *Environmental Chemistry* (Boston, MA: Willard Grant Press, 1972), 1.
170 Bernadette Bensaude-Vincent and Jonathan Simon, *Chemistry: The Impure Science* (London: Imperial College Press, 2008), 207. See Chapter 1.

thematic approach was on the chemistry of risk assessment.[171] That topic illustrates the importance of social concerns for DEC:[172]

> The risk assessment theme for the Washington meeting (...) may serve as a measure for doing this type of divisional programming. Local section programming on these issues was mentioned, as well as the difficulties faced by chemists in industry and government who wish to get involved in public issues. It was the general consensus that something positive should be done in terms of bringing together interested, responsible parties to address these issues.

As a follow-up to the *Montreal Protocol on Substances that Deplete the Ozone Layer* of 1987 DEC chairman Larry Keith mentioned the high interest in air toxics and the analysis of these compounds, and proposed for the Washington meeting a symposium titled Measurement of Airborne Compounds: Sampling, Analysis, and Data Interpretation.[173] The combination of public environmental concerns and the thematic programming of ACS National Meetings thus symbolizes the new context in which DEC operated. Another example is the ACS Committee on Environmental Improvement (CEI).[174] That committee asked DEC for a liaison, because it considered environmental issues and policies highly relevant to ACS members.[175] Moreover, CEI was involved in the communication with EPA and the public:[176]

> To further environmental risk assessment, management, and communication, President Gassman, Past President Callis, Joan B. Berkowitz (Chair of the ACS Committee on Environmental Improvement), and Allan M. Ford (Chair of the CEI Task Force on Air Toxics) visited with EPA Administrator William K. Reilly to explore ways in which the [American Chemical] Society and EPA could develop more effective interactions on issues of environmental concern.

---

[171] Minutes of the ACS Divisional Officers Conference/Program Coordination Conference, January 27-29, 1989, Dallas, TX.
[172] Minutes of the executive and open sessions of the ACS Council Committee on Divisional Activities, April 9, 1989, Dallas, TX, 8-9.
[173] Larry H. Keith, letter to Bob Jolly (Program Chairman of DEC), August 18, 1989. Keith proposes to organize this symposium jointly with the ACS Division of Analytical Chemistry.
[174] CEI was the successor of the ACS Subcommittee on Environmental Improvement.
[175] Minutes of the DEC Executive Committee Meeting, September 10, 1989, Miami, FL; Ray Garant, interview, July 22, 2010. Garant was liaison of CEI.
[176] ACS, *American Chemical Society Annual Report 1990* (Washington, DC: ACS, 1991), 4.

CEI was an important actor within ACS at the interface of chemistry, society and environmental policy, and DEC contributed to this committee.[177]

Relationships between EPA and DEC took place in various ways. For instance, DEC co-sponsored the so-called EPA Jekyll Island Conference.[178] The interaction between DEC and EPA is also illustrated by governing positions and executive positions. Richard Zepp (EPA) was Member-at-Large of DEC and Wayne Garrison (EPA) was the editor of *EnvirofACS*.[179] Furthermore, EPA officers contributed to symposia of DEC at the 200st ACS National Meeting in Washington, DC (1990), called 'Focus on Chemistry and Public Policy'.[180] At the Presidential Plenary Session of this meeting, Hank Habicht represented EPA with his lecture 'Understanding Risk Assessment', and Joseph Breen (EPA) lectured at several symposia of DEC.[181] EPA was also interested in collaborating with DEC on a new honorary award: an Environmental Chemist Award.[182] Such initiatives were consistent with DEC's policy goals, such as improving the awareness of environmental chemistry within the scientific community and the public at large.[183]

In sum, I conclude that the 1970s and 1980s witnessed a significant growth of environmental chemistry as a subfield within the chemical sciences (cf., Figure 2.3). Essentially, this growth kept pace with the increases in both public environmental concerns and, often, the public distrust in USA's policies on environmental control. This went hand in hand with a changing 'rationale' for the chemical sciences, as witnessed by a growing demand for both applied and basic environmental chemistry knowledge.[184] During those years a community of environmental chemists emerged acquainted with the analysis and understanding of minute amounts and pathways of both manmade and natural chemicals in

---

[177] For instance, Ford was Member-at-Large of DEC. See: list of DEC Officers and Executive Committee Members-1990, which is dated July 13, 1990.
[178] ACS, "200th ACS National Meeting Aug. 26-31," *Chemical & Engineering News* 68 (1991): 43-137.
[179] DEC's newsletter has been called *EnvirofACS*, See also: list of DEC Officers and Executive Committee Members-1990.
[180] S. Allen Heininger, "President's Welcome," in ACS, "200th ACS National Meeting".
[181] ACS, "200th ACS National Meeting," 101-103. Breen was involved in DEC before. See for example: Joseph J. Breen and Philip E. Robinson, ed., *Environmental Applications of Chemometrics* (Washington, DC: ACS, 1985). See also: "F. Henry Habicht: Biography," EPA, November 17, 2019, https://archive.epa.gov/epa/aboutepa/f-henry-habicht-ii.html.
[182] Minutes of the DEC Executive Committee Meeting, September 10, 1989, Miami, FL. For a list of Directors in 1989 see the Articles of Incorporation of DEC that complies with the District of Columbia Nonprofit Corporation Act. See also: DEC, *ACS Division of Environmental Chemistry Secretary's Report for the Year 1989: The Environmental Division's 76th Year* (Washington, DC: ACS, 1990).
[183] See for example: DEC, *ACS Division of Environmental Chemistry Secretary's Report for the Year 1990: The Environmental Division's 77th Year* (Washington, DC: ACS, 1991).
[184] Hessels and Van Lente, "Practical Applications as a Source of Credibility," 218.

environmental media. They employed interdisciplinary insights with respect to environmental quality and to understanding the environment.[185] EPA officers were among the dominant actors within the US community of environmental chemists. They were part and parcel of networks within DEC and that proved of great importance in the recognition of a new EPA policy that I will discuss now: pollution prevention.[186]

2.3 From Command-and-Control to Pollution Prevention, 1980s-2000

During the early 1980s, 'a troubled and emotional period for pollution control' according to Ruckelshaus, the command-and-control policy of EPA was criticized increasingly.[187] This was one of the reasons that so-called pollution prevention policies emerged, which dominated the environmental policies of the 1990s. That changing policy context became of great importance for the emergence of green chemistry in the USA in those years.

In the mid-1980s, Ruckelshaus, who just had been re-appointed as Administrator of EPA, argued that:[188]

> Imperfect as it is, I think the command and control process will work if we use it selectively and sensibly which I do not think we are doing.

Ruckelshaus argued that it would be better to use science-based policies in order to tackle environmental risks instead of exaggerated policy responses to public emotions. Moreover, in his view, the regulatory framework, as based on command-and-control, left too little choices for economic incentives to solve USA's environmental problems.[189] The views of Ruckelshaus illustrate that EPA's environmental policy of command-and-control was ready to change.

---

[185] For further reading, see also: CEI and ACS Subcommittee on Environmental Analytical Chemistry proposed guidelines in a frequently cited article: Daniel MacDougall et al., "Guidelines for Data Acquisition and Data Quality Evaluation in Environmental Chemistry," *Analytical Chemistry* 52 (1980): 2242-2249.

[186] For further reading: Bruno Latour, *Science in Action* (Cambridge, MA: Harvard University Press, 1987).

[187] William D. Ruckelshaus, "Science, Risk and Public Policy," *Science* 26 (1983): 1026-1028. See also: Annette T. Rosenblum, "Government Affairs," in *The American Chemical Society*, Reese, 64-74.

[188] Ruckelshaus, "Environmental Protection," 469. This time, Ruckelshaus was the successor to Gorsuch.

[189] Ibid., 468-469; Ruckelshaus, "Science, Risk," 1026-1028.

Already in 1984 US Congress adopted a modification of the command-and-control policy on the basis of the Hazardous and Solid Waste Amendments to the RCRA. The main purpose of these amendments was to 'reduce or eliminate, as expeditiously as possible, the amount of waste generated, including hazardous waste'.[190] Environmental policy thus increasingly shifted to the prevention of environmental problems. This was reported in 1986 by the EPA explicitly to the US Congress.[191]

By 1989 pollution prevention, instead of command-and-control, had become EPA's policy paradigm as the option of first choice.[192] In line with the legal initiatives by EPA, mentioned above, that new approach was not intended to soften existing environmental regulations and goals. But for economic, environmental and social reasons EPA now started to cooperate increasingly with the chemical industry in developing new processes. Economically, because pollution prevention had to reduce the costs of management, handling, control and disposal of toxic chemicals. Environmentally, because the release of pollutants to the environment should be reduced or eliminated at the source. Socially, last but not least, because 'pollution prevention [was] consistent with the public's "right-to-know" and "right-to-know more" laws, and with increased public scrutiny of industrial practices', according to the EPA officers Joseph Breen and Michael Dellarco.[193]

Internationally, the idea of pollution prevention was not new. Some individual firms had started already in the 1970s with the introduction of prevention measures in order to reduce the costs of handling environmental wastes.[194] Influenced by global environmental politics, the USA had signed the *1979 Convention on Long-Range Transboundary Air Pollution* (*LRTAP*), which entered into force in 1983.[195] *LRTAP* has been urging for

---

[190] "Solid Waste Disposal Act," in *Compilation of Selected Acts*, Committee on Commerce U.S. House of Representatives, 455-613.

[191] Stephan and Atcheson, "The EPA's Approach," 53-58.

[192] Ibid. Stephan served as senior engineering advisor at EPA's Risk Reduction Engineering Laboratory in Cincinnati, OH, and Atcheson was director of the Implementation and Outreach Staff, Pollution Prevention Office at EPA in Washington, DC. See also Dyer (EPA) who described the assistance of EPA to chemical industry: Norman E. Dyer, "Technology Transfer/Pollution Prevention," *Journal of Hazardous Materials* 22 (1989): 264-265.

[193] Joseph J. Breen and Michael J. Dellarco, "Pollution Prevention: the New Environmental Ethic," in *Pollution Prevention in Industrial Processes: the Role of Process Analytical Chemistry*, ed. Joseph J. Breen and Michael J. Dellarco (Washington, DC: ACS, 1992), 2-12.

[194] Royston lists pollution prevention actions of chemical companies: Michael G. Royston, "Making Pollution Prevention Pay," *Harvard Business Review* 58 (1980): 6-27.

[195] *LRTAP* was signed by 35 countries. See: United Nations, *Convention on Long-Range Transboundary Air Pollution* (Geneva: United Nations, 1979). In United Nations Treaty Collection; available from https://treaties.un.org/Pages/ViewDetails.aspx?src=TREATY&mtdsg_no=XXVII-1&chapter=27&clang=_en, accessed May 22, 2022; Pamela S. Chasek, David L. Downie and Janet W. Brown, *Global Environmental Politics* (Philadelphia, PA: Westview Press, 2010), 121-129; Willemijn

international cooperation and provided some 'fundamental principles' concerning environmental policy:[196]

> The Contracting Parties, taking due account of the facts and problems involved, are determined to protect man and his environment against air pollution and shall endeavour to limit and, as far as possible, gradually reduce and prevent air pollution including long-range transboundary air pollution.

Based on a consensus seeking policy style, the implementation of these fundamental principles was laid down in binding protocols with some countries that often took the lead, e.g., Sweden and the Netherlands. These countries aimed by the mid-1980s for a 30 % reduction of emissions of pollutants, such as SO_2, in order to combat acid rain and to improve the environmental quality of air. But the USA often refused to follow that lead:[197]

> In the initial negotiations, a group of veto states, usually led by the United Kingdom, Germany, and/or the United States, either refused to join particular protocols or demanded that they be watered down.

This marks the ambivalence during the mid-1980s of the USA towards *LRTAP*. On the one hand there were rules and norms prescribed in international protocols, that conflicted with USA's political culture of sovereignty and its economic interests. On the other hand, the holistic principles of *LRTAP* were recognized and shaped by the USA in its own way. For example, the CAA Amendments of 1990 promoted tightened environmental policies to combat acid rain and the reduction of air pollution at source.[198]

As a result of these developments, pollution prevention became of increasing importance within EPA. In 1988, EPA 'created a new Pollution Prevention Office (PPO) within its Office of Policy Planning and Evaluation and established a Waste Minimization

---

Tuinstra, Leen Hordijk and Carolien Kroeze, "Moving Boundaries in Transboundary Air Pollution Co-production of Science and Policy under the Convention on Long Range Transboundary Air Pollution," *Global Environmental Change* 16 (2006): 349-363; Adam Byrne, "The 1979 Convention on Long-Range Transboundary Air Pollution: Assessing its Effectiveness as a Multilateral Environmental Regime after 35 Years," *Transnational Environmental Law* 4 (2014): 37-67.

[196] United Nations, *Convention on Long-Range Transboundary Air Pollution.*

[197] Chasek, Downie and Brown, *Global Environmental*, 128. See also: Byrne, "The 1979 Convention," 45.

[198] "Clean Air Act," in *Compilation of Selected Acts*, Committee on Commerce U.S. House of Representatives, 7 and 324-374. See also: Chasek, Downie and Brown, *Global Environmental*, 125.

Branch in its Risk Reduction Engineering Laboratory in Cincinnati, OH'.[199] PPO chose four approaches to aim for pollution prevention. Firstly, EPA published a Pollution Prevention Policy Statement in the January 26, 1989 Federal Register. Secondly, EPA published existing statutes and policies that were reviewed and modified from a pollution prevention perspective. Thirdly, PPO searched for financial incentives for pollution prevention initiatives and for opportunities for increasing regulatory flexibility. And in the fourth place, the PPO cooperated with other offices within EPA 'to assure that the federal government's own house [was] in order'.[200] Together, these measures led to the awarding of grants in order to support pollution prevention programs and to stimulate the cooperation between government and industries in particular. PPO was created even before the concept of pollution prevention was politically formalized and therefore illustrates the urge for pollution prevention. But soon that new policy would be implemented by law.[201]

In 1990 US Congress passed the Pollution Prevention Act (PPA), under the Republican Administration of President George Bush.[202] This occurred in a difficult economic period (cf., Figure 2.1).[203] The US Congress concluded that the USA produced 'millions of tons of pollution' and that controlling this pollution entailed approximately 'tens of billions of dollars per year'.[204] That conclusion underlined the environmental and economic urge to adopt the policy of pollution prevention. EPA had to take a leading role in this new policy field.[205]

The proposed activities of EPA included facilitating 'the adoption of source reduction techniques by business'.[206] Extra money was therefore budgeted within EPA, aiming for new pollution prevention programs. To achieve that goal, institutional barriers had to be overcome, because EPA was to a certain extent still organized 'along single environmental medium lines'.[207] Therefore, in 1991 EPA issued its Pollution Prevention Strategy.[208] According to this strategy, pollution prevention had to become the bottom-line of all EPA's operations, including 'conducting research on industrial pollution prevention and waste minimization

---

[199] Stephan and Atcheson, "The EPA's Approach," 53-58.
[200] Ibid.
[201] Ibid.
[202] "Pollution Prevention Act of 1990," in *Compilation of Selected Acts*, Committee on Commerce U.S. House of Representatives, 615-623. See also: Breen and Dellarco, "Pollution Prevention," 2-12.
[203] See also: Norton et al., *A People and a Nation*, 371.
[204] "Pollution Prevention Act," 617.
[205] Ibid., 618-619.
[206] Ibid., 619.
[207] Breen and Dellarco, "Pollution Prevention," 9.
[208] Ibid., 2-12.

technologies'.[209] EPA emphasized the importance of technical assistance to chemical industries, as this would improve the implementation of pollution prevention programs.[210] This would help solving environmental problems, but Breen and Dellarco also proposed a non-environmental dimension: the public at large:[211]

> Establishing cash awards for EPA facilities and individuals who devise policies or actions to promote pollution prevention; publicly commending and publicizing industrial facility pollution success stories.

These actions were part of the law, because PPA prescribed the establishment of 'an annual award program to recognize a company or companies which operate outstanding or innovative source reduction programs'.[212]

I conclude that the PPA embodied a principal change in environmental policy from command-and-control to prevention. The new law also provided a changing 'rationale' and 'conditions' for the chemical sciences and thus new opportunities for the development of scientific programs with societal relevance.[213] These new pollution prevention programs were almost instantaneously developed within the new policy context, with scientists at EPA taking the lead.

2.4 EPA and DEC: the Start of Green Chemistry

During the early 1990s, the idea of pollution prevention was extensively discussed at divisional symposia of DEC as were related activities such as new methods in analytical chemistry. Most of these symposia were chaired by EPA officers, Breen and Dellarco in particular.[214] These symposia contributed to placing pollution prevention seriously on the scientific agenda. Within the ACS, this resulted in the launching of the CEI Pollution

---

209 Ibid.
210 Ibid.; Stephan and Atcheson, "The EPA's Approach," 53-58.
211 Breen and Dellarco, "Pollution Prevention," 9.
212 "Pollution Prevention Act," 619.
213 Hessels and Van Lente, "Practical Applications as a Source of Credibility," 218.
214 ACS, "201st National Meeting Atlanta," *Chemical & Engineering News* 69 (1991): 65-67; Minutes of the DEC Executive Committee Meeting, August 25, 1991, New York, NY. See also: ACS, "Technical Program & Exposition. 4th Chemical Congress of North America & 202nd ACS National Meeting," *Chemical & Engineering News* 69 (1991): 53-123. In San Francisco, at the 203rd ACS National Meeting, there were also pollution prevention lectures at symposia of DEC. See: ACS, "Technical Program: April 5-10, 1992," *Chemical & Engineering News* 70 (1992): 66.

Prevention Subcommittee and to more pollution prevention symposia at ACS National Meetings.[215] In 1992 a new dimension was added at the 204th ACS National Meeting in Washington, DC. Under leadership of Breen, DEC then organized a pollution prevention symposium with a lecture by Roger Garrett (EPA) about 'pollution prevention in synthetic chemical design'.[216] This shows new initiatives aimed at connecting pollution prevention to the design stage of chemicals, instead of methods to measure pollution afterwards, as we will see below.

In 1993, the ACS informed DEC about the article written by Ivan Amato: "The Slow Birth of Green Chemistry".[217] This was soon discussed within the Executive Committee of DEC.[218] The term 'green chemistry' used by Amato was then rather new and hardly used by USA chemists.[219] Also Amato did not give a very precise meaning to this term, and – despite the title of his article – used mainly the term 'environmental chemistry' in the article itself.[220] After discussing the skills of environmental chemists in chemical analysis, following the end-of-pipe approach, Amato explained the new approach developed by EPA. He discussed the new pollution prevention program of EPA, called Design for the Environment (DfE): the Chemical Design Project.[221] The aim of that Chemical Design Project was that chemistry should become more environmentally benign. Some chemists had already received research funding for the development of chemical processes that prevented pollution. Amato emphasized that this approach was new, and he called it a new environmental chemistry.

Within EPA, Joseph Breen was an advocate of that new approach.[222] He teamed up with Paul Anastas, a new actor at the stage at EPA. Anastas had studied chemistry at the

---

[215] Minutes of the DEC Executive Committee Meeting, August 25, 1991, New York, NY; Minutes of the DEC Executive Committee Meeting, April 5, 1992, San Francisco, CA.
[216] ACS, "Technical Program & Exposition: 204th ACS National Meeting and Exposition Program, Washington, D.C., August 23-28, 1992," *Chemical & Engineering News* 70 (1992): 63. This meeting took place in the same city as EPA's headquarter.
[217] Cathy A. Nelson (ACS Committee on Professional Training), letter to Alan Elzerman (DEC), February 22, 1993. Nelson attached to this letter: Ivan Amato, "The Slow Birth of Green Chemistry," *Science* 259 (1993): 1538-1541.
[218] Minutes of the DEC Executive Committee Meeting, August 22, 1993, Chicago, IL, 8. Owing to its novelty, the term "green chemistry" was placed between quotation marks in these minutes.
[219] Most likely, an Irish man seems to be the first who used the term green chemistry in a context of environmental concerns: Clive Cathcart, "Green Chemistry in the Emerald Isle," *Chemistry & Industry* 21 (1990): 684-687.
[220] Amato, "The Slow Birth," 1538-1541. Just like Amato, Cathcart used the term "green chemistry" without giving (chemical) meaning to this term.
[221] Often, this project was merely named Chemical Design Project within EPA (and thus also in this thesis).
[222] In 1993, Breen became also Program Area Chair of pollution prevention within DEC. See: Minutes of the DEC Long Range Program Planning Committee Meeting, Denver, CO, March 29, 1993.

University of Massachusetts, where he received his BSc in chemistry. In 1989, he completed his PhD in synthetic organic chemistry at Brandeis University.[223] By then he was not yet focused on pollution prevention in the development of organic syntheses, nor with environmental chemistry. Anastas left the laboratory and was appointed within the EPA Office of Pollution Prevention and Toxics (OPPT). Within that office he got the opportunity to develop and coordinate the project to which Amato referred. A few years later, he thanked Garrett 'for putting his faith in a young chemist to develop the program'.[224]

The goals of the new project were:[225]

> EPA's Chemical Design Project aims to change the way organic chemists approach the design of synthetic pathways for chemical production. EPA is encouraging consideration of alternative synthetic pathways through: (1) grant awards to academic institutions; (2) discussion at national symposia; and (3) use of computer programs which assist in the design of chemical pathways.

The focus of the project was on organic chemists who traditionally focused on yield and low cost of syntheses, but ignored pollution prevention.[226] OPPT wanted to fund organic chemists who designed alternative synthetic pathways that could 'avoid many of the environmental problems and liabilities chemical producers face before they arise'.[227] These new pathways would have economic benefits due to the lower costs of both waste disposal and compliance. There were also benefits for both the environment and human health, because of the fact that the generation of toxic by-products and the use of toxic substances were both reduced. OPPT followed the legal mandate of PPA. Based on that new act, extra money was indeed budgeted

---

[223] John C. Warner, Paul T. Anastas and Jean-Pierre Anselme, "The Wittig Reaction in the Undergraduate Organic Laboratory," *Journal of Chemical Education* 62 (1985): 346; Paul T. Anastas and Robert Stevenson, "Synthesis of Natural Lignan Arylnapthalene Lactones, Deurinol and Retrochinensin," *Journal of Natural Products* 54 (1991): 1687-1691; Paul T. Anastas et al., "Condensation of Ethyl Mandelate with a Cinnamate Ester: Bicyclocarbolactone Formation," *Journal of Chemical Research (S)* 3 (1991): 54-55.
[224] Paul T. Anastas, "Benign by Design Chemistry," in *Benign by Design: Alternative Synthetic Design for Pollution Prevention*, ed. Paul T. Anastas and Carol A. Farris (Washington, DC: ACS, 1994), 2-22.
[225] Paul T. Anastas, "Design for the Environment: Chemical Design Project," (1993): 1-2 [Internal EPA communication]. See also: Paul T. Anastas, "Design for the Environment: Chemical Design Project; Changing the Way Chemists Think about Chemicals," (AL: probably early 1993): 1-2 [Internal EPA communication].
[226] Anastas, "Design for the Environment," 1-2.
[227] Ibid., 1.

for EPA to develop new programs aiming at pollution prevention. In 1992 EPA awarded the first research grants in these topics.[228]

In 1993, the OPPT signed a Memorandum of Understanding on pollution prevention with the Chemistry Division of the National Science Foundation (NSF).[229] Under leadership of Kenneth Hancock and in concert with the Council for Chemical Research, the NSF had launched the Environmentally Benign Chemical Synthesis and Processing Program. The aims of this funding program overlapped with those of the Chemical Design Project. This resulted in a Memorandum of Understanding that allowed EPA to grant approximately two million dollars extra for research projects on alternative synthetic pathways.[230]

The research projects executed under these funding schemes were soon discussed at two DEC symposia at the 206th ACS National Meeting. They were organized by Carol Farris (EPA) and Anastas.[231] These symposia were titled Alternative Synthetic Design for Pollution Prevention.[232] The chemists who received the first grants of the Chemical Design Project lectured at these symposia.[233] Other lectures were given by Farris and Anastas on EPA's pollution prevention policies. Also a few lectures were given by industrial chemists. In 1994 Farris and Anastas published the results of these Alternative Synthetic Design for Pollution Prevention symposia in the book *Benign by Design: Alternative Synthetic Design for Pollution Prevention.*[234] The contributions to that book had not only been funded by EPA and NSF, but for at least 50 % also by chemical companies.

---

228 "Pollution Prevention Act," 623; Carol A. Farris (EPA), interview, July 26, 2010.
229 Kenneth G. Hancock and Margaret A. Cavanaugh, "Environmentally Benign Chemical Synthesis and Processing for the Economy and the Environment," in *Benign by Design*, Anastas and Farris, 29.
230 Anastas, "Design for the Environment," 1; Anastas, "Benign by Design," 18.
231 ACS, "Technical Program & Exposition: 206th ACS National Meeting August 22-27," *Chemical & Engineering News* 71 (1993): 84.
232 Ibid. The ACS Division of Organic Chemistry was the co-sponsor. Here, the symposia were titled "Alternate Synthetic Design for Pollution Prevention". In 1994 the same symposia were titled "Alternative Synthetic Design for Pollution Prevention", see: Anastas and Farris, *Benign by Design.*
233 The recipients were: Gary Epling (University of Connecticut) with "Replacing Heavy Metal Catalysts"; Orville Chapman (UCLA) with "New Synthesis of Styrene"; George A. Krauss (Iowa State University) with "Visible Light to Replace the Friedel-Crafts Reaction"; John W. Frost (Purdue University) with "Replacing Benzene with Simple Sugars"; James M. Tanko (Virginia Polytechnic Institute and State University) with "Alternative for Toxic Solvents". See heretofore: Anastas, "Design for the Environment," 2; Anastas, "Design for the Environment: Chemical Design Project; Changing the Way Chemists Think about Chemicals," 2. Barry B. Snider (Brandeis University) was not awarded, but also lectured about his project on the elimination of tin based catalysts.
234 Anastas and Farris, *Benign by Design.*

The Chemical Design Project and, especially, the book *Benign by Design* put the new term 'benign by design chemistry' firmly on the map.[235] It was defined as follows:[236]

> Benign by design chemistry is synthetic chemistry designed to use and generate fewer hazardous substances. The ultimate goal of benign chemistry is to develop and institute alternative syntheses for important industrial chemicals in order to prevent environmental pollution.

Just like Garrett had done earlier, Anastas and Farris connected "benign by design chemistry" to pollution prevention and synthetic chemistry.[237] They also argued that the authors of the papers in their book were 'pioneers' in the field of benign by design chemistry and that their symposium 'was the first one to explore this new area of research in synthetic chemistry'.[238]

The new term benign by design chemistry was frequently used in the book, but the term "green chemistry", mentioned by Amato, only rarely. One of the few exceptions were James Tanko et al., who also cited Amato, in their paper on carbon dioxide as an 'environmentally-benign alternative solvent'.[239] They stated:[240]

> The successful use of CO_2 as a solvent represents an important early step in the growth and development of "green chemistry".

Although there were clear signs that the term "green chemistry" had entered DEC and EPA, the term was not frequently used by chemists. It can therefore be concluded that EPA officers mostly favored the term benign by design chemistry (or benign chemistry), and via their networks these terms and the ideas on pollution prevention were also spread within DEC.

---

235 Paul T. Anastas and Carol A. Farris, preface to *Benign by Design*, Anastas and Farris, IX-X.
236 Ibid., IX.
237 Farris, interview, July 26, 2010. She argued that Garrett started with the idea of alternative synthetic pathways and pollution prevention.
238 Anastas and Farris, preface to *Benign by Design*, Anastas and Farris, IX-X.
239 James M. Tanko, Joseph F. Blackert and Mitra Sadeghipour, "Supercritical Carbon Dioxide as a Medium for Conducting Free-radical Reactions," in *Benign by Design*, Anastas and Farris, 98-113.
240 Ibid., 112. Hancock suddenly passed away in September 1993. Anastas and Farris praised him in their obituary for his 'commitment to green chemistry and pollution prevention'. See: Paul T. Anastas and Carol A. Farris, "In Memoriam," in *Benign by Design*, Anastas and Farris, XI. In a posthumous publication by Hancock and Margaret Cavanaugh (NSF) in *Benign by Design*, which discussed the NSF's Environmentally Benign Chemical Synthesis and Processing Program, the terms 'benign chemistry' and 'environmental chemistry' were used. See: Hancock and Cavanaugh, "Environmentally Benign Chemical Synthesis," 23-30.

Anastas played a key-role in spreading the idea of benign by design chemistry at ACS National Meetings.[241] For instance, a large symposium was sponsored by EPA and DEC in 1994, on Design for the Environment: The Environmental Paradigm for the 21st Century, organized by Breen and Anastas.[242] The aim was that this symposium would 'afford an opportunity for analytical and synthetic chemists, chemical engineers and economists, industrial and environmental scientists, management, and policy makers to report progress and exchange ideas on implementation of pollution prevention as an integral part of our national environmental and economic policies'.[243] In their view the 'role of the chemist and Benign by Design Chemistry is central to the success of the DfE approach.' [244] Twenty DfE symposia took place in 1994 at the 208th ACS National Meeting in Washington, DC.[245] In comparison with the former pollution prevention efforts within DEC, this time the effort was significantly larger and included also international speakers. Anastas himself gave a lecture Benign by Design Chemistry: Pollution Prevention through Synthetic Design.[246]

Although during the symposia Anastas mainly used the term "benign by design chemistry", during the publication of their results a significant move was made. Together with EPA officer Tracy Williamson (OPPT), Anastas collected the presentations of the DfE symposia in the book *Green Chemistry: Designing Chemistry for the Environment*, which was published in 1996.[247] In the first chapter of that book, "Green Chemistry: An Overview", the EPA approaches were presented with respect to pollution prevention and the design of synthetic processes for the environment.[248] Most authors in *Green Chemistry: Designing* did not use the term green chemistry though, and the only two authors who did – James Huheey

[241] According to Garant, Anastas is a creative man: Garant, interview, July 22, 2010. For example, Anastas lectured about benign by design chemistry and education: ACS, "207th ACS National Meeting Technical Program & Exposition March 13-17," *Chemical & Engineering News* 72 (1994): 29-133. Next to the industry and the academic arena, PPA obliged EPA to focus pollution prevention programs on education: "Pollution Prevention Act," 619.
[242] Joseph J. Breen and Paul T. Anastas, "Design for the Environment: the Environmental Paradigm for the 21st century," *Chemtech* 24 (1994): 6-7. Joseph Breen, fax of this article to Gordon Bellen, July 26, 1994. See also: Mary Walker in Minutes of the DEC Executive Committee Meeting, March 13, 1994, San Diego, CA.
[243] Breen and Anastas, "Design for the Environment: the Environmental Paradigm," 6.
[244] Ibid.
[245] ACS, "208th National Meeting August 21-25, 1994," *Chemical & Engineering News* 72 (1994): 53-143. The DfE symposia were organized in memory of Hancock. Breen and Anastas were in the EPA office at Washington, DC.
[246] Ibid., 83.
[247] Paul T. Anastas and Tracy C. Williamson, ed., *Green Chemistry: Designing Chemistry for the Environment* (Washington, DC: ACS, 1996).
[248] Compare: Anastas, "Benign by Design" with Paul T. Anastas and Tracy C. Williamson, "Green Chemistry: An Overview," in *Green Chemistry: Designing*, Anastas and Williamson, 1-17.

(University of Maryland) and Milagros Simmons (University of Michigan) – did not refer to publications of Anastas on green chemistry and benign by design chemistry.[249] Nevertheless, the shift in terminology by Anastas and Williamson was not accidental, as will be discussed in the next sections.[250]

2.5 Anticipation of Polarization: *Reinventing Environmental Regulation*

Previous sections discussed the role of EPA in, what I call, the "triangle of distrust" between industry, government and society. The associated tensions and challenges in USA's contentious civic epistemology became subject to political discourse in the early 1990s, which in turn impacted EPA's operations.[251] It is from that changing policy culture that DfE developed, which subsequently gave rise to the term green chemistry and the practices associated with that. The political recognition of the new term and its practices was of great significant, as I will show.

The years around 1990 were a turning point. As a result of the 1990 CAA Amendments (with its tough air quality standards)[252] and the PPA, dozens of new programs were developed within EPA that principally differed from traditional policies, and which took all three angles of the "triangle of distrust" into account. Several programs were developed in cooperation between EPA and private companies, together with an increased public transparency.[253] An example is the development of trade emissions programs that should improve pollution prevention innovations at the plant level. The programs developed by EPA were also intended though to get a fuller participation by society. Examples are the Project

---

[249] James E. Huheey, "Incorporating Environmental Issues into the Inorganic Curriculum," in *Green Chemistry: Designing*, Anastas and Williamson, 232-238; Milagros S. Simmons, "The Role of Catalysts in Environmentally Benign Synthesis of Chemicals," in *Green Chemistry: Designing*, Anastas and Williamson, 116-130.

[250] Jody A. Roberts, "Creating Green Chemistry: Discursive Strategies of a Scientific Movement," (PhD diss., Virginia Polytechnic Institute and State University, 2005). In VirginiaTechWorks; available from https://vtechworks.lib.vt.edu/handle/10919/27529, accessed May 22, 2022; Jurgen Exner (DEC), (phone) interview, July 23, 2010. In the period of terminology shift, Exner was Chair-Elect of DEC.

[251] Bill Clinton and Al Gore, *Reinventing Environmental Regulation* (Washington, DC: National Performance Review, 1995), 1-61. See also: Rena I. Steinzor, "Reinventing Environmental Regulation: the Dangerous Journey from Command to Self-control," *Harvard Environmental Law Review* 22 (1998): 103-202. See for the civic epistemology concept Chapter 1 and, see: Jasanoff, *Designs on Nature*, 247-271.

[252] "Clean Air Act," 1-451.

[253] Steinzor, "Reinventing Environmental Regulation," 103-202. See also: Hoffman, *From Heresy to Dogma*, 111-113. Hoffman discusses voluntary programs, e.g., 33/50 and Green Lights.

XL and the Common Sense Initiative, or the DfE program.[254] DfE was thus developed in a changing political environment that allowed EPA, in line with his legal mandates, to reduce social polarization.

On March 16, 1995, the Democratic Administration of Bill Clinton/Al Gore announced the *Reinventing Environmental Regulation* program.[255] While much environmental progress had been made, Clinton and Gore argued that time had come to change the principles of USA's environmental policy.[256] Their government introduced the 'cleaner, cheaper and smarter' slogan in its program for sustainable development.[257] Explicitly this new program intended to steer away from legal and social polarization. Consequently, new policy principles (Table 2.1) were formulated that were also new with respect to the style of policy making in USA's political culture, according to the *Reinventing Environmental Regulation* document. On the one hand, these principles accounted for environmental goals and rights with respect to economical perspectives. But on the other hand, the desired style of policy making was explicitly consensus seeking (cf., Table 2.1; principles 6, 8 and 9) instead of the traditional strict focus on law and order, thereby respecting individual environmental rights (cf., Table 2.1; principle 10).

No.	*Principle*
1	Protecting public health and the environment are important national goals, and individuals, businesses and government must take responsibility for the impact of their actions.
2	Regulation must be designed to achieve environmental goals in a manner that minimizes costs to individuals, businesses, and other levels of government.
3	Environmental regulations must be performance-based, providing maximum flexibility in the means of achieving our environmental goals, but requiring accountability for the results.
4	Preventing pollution, not just controlling or cleaning it up, is preferred.
5	Market incentives should be used to achieve environmental goals, whenever appropriate.

[254] For further reading on Project XL and the Commons Sense Initiative, see: Steinzor, "Reinventing Environmental Regulation," 103-202.
[255] Clinton and Gore, *Reinventing*.
[256] This is consistent with Spitz his claims concerning the US chemical industry: Spitz, "The Chemical Industry and the Environment," 230-233.
[257] Clinton and Gore, *Reinventing*, 5.

6	Environmental regulation should be based on the best science and economics, subject to expert and public scrutiny, and grounded in values Americans share.
7	Government regulations must be understandable to those who are affected by them.
8	Decision making should be collaborative, not adversarial, and decision makers must inform and involve those who must live with the decisions.
9	Federal, state, tribal and local governments must work as partners to achieve common environmental goals, with non-federal partners taking the lead when appropriate.
10	No citizen should be subjected to unjust or disproportionate environmental impacts.

Table 2.1: The ten environmental policy principles of *Reinventing*.[258]

Reinventing emphasized the policy importance of the pollution prevention idea and therefore also the DfE program of EPA (cf., Table 2.1; principle 4).[259] The Clinton/Gore Administration believed that pollution prevention could lead to economic and environmental benefits. Moreover, following the legal framework of PPA, the EPA was mandated to 'establish an annual award program to recognize a company or companies which operate outstanding or innovative source reduction programs'.[260] This provided, given the "triangle of distrust", a new dimension to the DfE program as illustrated by one of the '25 High Priority Actions' presented in *Reinventing*: the Design for the Environment – Green Chemistry Challenge. So, in this particular political context, DfE became associated with the term green chemistry. It was formulated that the goal of DfE was to 'promote pollution prevention and industrial ecology through a new EPA Design for the Environment partnership with the chemical industry'.[261] Key action of DfE became an EPA award system that would lead to pollution prevention measures by companies, such as designing more safely compounds.

Economic and environmental considerations for this new EPA policy were important for the Clinton/Gore Administration, including the generation of positive publicity:[262]

[258] Ibid., 6.
[259] Ibid., 10. The Clinton/Gore Administration introduced the so-called Sustainable Development Challenge Grants.
[260] "Pollution Prevention Act," 619.
[261] Clinton and Gore, *Reinventing*, 42. See Chapter 5 for further discussion on industrial ecology.
[262] Ibid. According to Farris, Anastas was the man who realized that the awards became Presidential by making use of the right momentum, i.e., the preparation of *Reinventing*: Farris, interview, July 26, 2010.

Publicity associated with the awards program, coupled with financial prizes provided by the chemical industry, can provide a strong incentive for broad industry cooperation.

The publicity around the Presidential Green Chemistry Challenge Awards, as the program was called, was to demonstrate good chemical practices in order to 'build trust among [the] traditional adversaries' society and the chemical industry, and chemistry at large.[263] That also explains the interest of the ACS Committee on Environmental Improvement (CEI) to cooperate with EPA in the organization of the Presidential Green Chemistry Challenge Awards.[264] As pointed out in Chapter 1 of this thesis, CEI and the ACS at large saw a strong relationship between environmental concerns and the negative public image of chemistry.[265] Therefore, one of the mission statements of the CEI Pollution Prevention Subcommittee was public outreach. Supporting the Presidential Green Chemistry Challenge Awards was perfect for that mission.[266]

The term green chemistry became in this way a rhetorical instrument in the dialogue between society, government and chemical industry, and chemistry at large. In the context of the changing environmental policy principles of the Clinton/Gore Administration, it even received political recognition in the form of a Presidential awards program. Chemists within the ACS saw this as an instrument to improve the image of chemistry. This explains why the ACS easily embraced the term green chemistry, given its worries about the social image of chemistry.[267] We have to be aware though that at the time of the start of the Presidential Green Chemistry Challenge Awards (1995) and the publication of *Green Chemistry: Designing* (1996), the term green chemistry was not widely embraced by chemists in

---

[263] Clinton and Gore, *Reinventing*, 1-3.
[264] Garant claimed that CEI was involved in the organization of the Presidential Green Chemistry Challenge Awards. Garant, interview, July 22, 2010.
[265] See also for example: Minutes of the executive and open sessions of the ACS Council Committee on Divisional Activities, April 9, 1989, Dallas, TX.
[266] CEI, "Committee on Environmental Improvement," *EnvirofACS* 40 (1992): 3-4. For ACS, green chemistry was a tool to communicate environmental improvements to the public at large and not to hide something: Garant, interview, July 22, 2010; Exner, interview, July 23, 2010. See with respect to 'green labeling' also: CEI, "Committee on Environmental Improvement," 3.
[267] This is consistent with the opinion of Exner, who argued that the shift in terminology from benign by design chemistry into green chemistry was 'to raise the image': Exner, interview, July 23, 2010. Clark (University of York), who contributed to a green chemistry movement in the UK and abroad, argued that EPA alleged a negative public image due to the operations of chemists: James H. Clark, (phone) interview, December 7, 2010. See also Chapter 3.

academia and industry. At that time there was no strong support in scientific circles of the green chemistry rhetoric, as we will see below.[268]

2.6 Beyond Environmental Chemistry: the Institutionalization of Green Chemistry

During the second part of the 1990s EPA and DEC continued their efforts to support the growth of green chemistry. Jurgen Exner (Chair of DEC), for instance, confirmed the support of DEC as co-sponsor to green chemistry conferences, e.g., the 1997 Green Chemistry & Engineering Conference: Implementing Vision 2020 for the Environment.[269] So, the organization of green chemistry activities thus continued within DEC, but this also took place more independently from DEC at the highest ACS levels. In what follows, I will argue that green chemistry received institutional recognition within ACS especially at its top level.

As demonstrated above, EPA was strongly involved in the Presidential Green Chemistry Challenge Awards program. This environmental protection agency defined green chemistry by the mid-1990s as 'the use of chemical principles and methodologies for source reduction', which is 'the highest tier of the risk management hierarchy as described in the Pollution Prevention Act of 1990'.[270] This definition played a role in calls for nominations for the new awards, which concerned topics such as 'alternative synthetic pathways' or 'design of chemicals'.[271] *EnvirofACS* subsequently reported that the first awards had to be presented at 1996 Earth Day and that ACS would organize the ceremony.[272]

---

[268] Farris argued that chemists were restrained in associating chemicals and chemical processes with green chemistry, but EPA did: Farris, interview, July 26, 2010.
[269] Jurgen Exner (Chair of DEC), letter to Margaret Carey, December 2, 1996; Minutes of the DEC Executive Committee Meeting, September 7, 1997, Las Vegas, NV. This conference was co-sponsored by the ACS Division of Industrial and Engineering Chemistry and within this division a small interest group on green chemistry developed. See: David E. Gushee, "Industrial and Engineering Chemistry," in *The American Chemical Society*, Reese, 118-119; Susan Turner (CEI) and Margaret Carey (ACS Department of Government Relations and Science Policy), letter to Paul Anastas, Alan Ford and the members of CEI, members of ACS Committee on Chemistry and Public Affairs, Chair and Secretary of DEC and Chair and Co-Chair of ACS Division of Industrial and Engineering Chemistry, December 13, 1996; Anonymous, "Green Chemistry & Engineering," *The Capitol Connection* 1 (1997): 2.
[270] EPA, "The President's Green Chemistry Challenge Awards Program: Nomination Package," (AL: probably Summer of 1995): 1 [Internal EPA communication].
[271] Ibid.
[272] Anonymous, "Green Chemistry Award," *EnvirofACS* 43 (1995): 9. The new awards were announced in several magazines. See for example: Judy Stringer, "Green Chemistry – EPA Launches Challenge," *Chemical Week* 157 (1995): 31.

The ACS voluntary cooperated in the Presidential Green Chemistry Challenge Awards program, together with EPA and some industrial corporations.[273] The nominations for the new awards were divided along the lines of industrial and academic interests. In 1996, for the category "Small Business" 8 nominations were received; for "Industry and Government" 54; for "Academia" 10. Considering the nominations for the latter, Hudlicky (University of Florida) and James Hendrickson (Brandeis University) were among the few academic nominees and who had both published in *Green Chemistry: Designing*, but then not yet employed the term green chemistry.[274] Considering the other nominees, none of them had published in *Benign by Design*, or in *Green Chemistry: Designing*.[275] Considering the categories Small Business and Industry & Government, approximately 40 different companies, employing different chemical methodologies, submitted one or more nominations and some of them jointly with governmental agencies, such as the US Department of Energy. The majority of these companies was not represented with a publication or financial support in *Benign by Design*, or in *Green Chemistry: Designing*.[276] So obviously, by the mid-1990s the term green chemistry was embraced by some academic chemists and companies at the interface of science, policy and society, but not yet used in wider circles.[277] In the meantime those chemists and companies had not played a significant role in the early green chemistry and pollution prevention symposia organized by DEC and EPA.

In line with the objectives published in *Reinventing*, the awards were sponsored by EPA and stakeholders from the chemical industry. The role of the ACS was to judge the entries of the nominees.[278] That the ACS involved was reported in its 1996 annual report:[279]

> An ongoing national effort to improve the environment is underway, and ACS is playing an important role. In 1996, we participated in the first [Presidential] Green Chemistry Challenge

[273] EPA, "The Presidential Green Chemistry Challenge Awards Program: Listing of 1996 Award Entries," Spring 1996 [Internal EPA communication].
[274] T. Hudlicky, "Environmentally Efficient Management of Aromatic Compounds: Economic Perspectives of Biocatalytic Conversion of Aromatics to Optically Pure Synthons for the Pharmaceutical Industry," in *Green Chemistry: Designing*, Anastas and Williamson, 180-197; James B. Hendrickson, "Teaching Alternative Syntheses: The SYNGEN Program," in *Green Chemistry: Designing*, Anastas and Williamson, 214-231.
[275] Anastas and Williamson, *Green Chemistry: Designing*; Anastas and Farris, *Benign by Design*.
[276] Exceptions in this respect are for example DuPont Company, Dow Chemical Company and Monsanto Company.
[277] See also: Roberts, "Creating Green Chemistry," 138-139.
[278] Paul T. Anastas and Tracy C. Williamson, "Design for the Environment: The Green Chemistry Challenge, Quick Reference Fact Sheet," July 1996 [Internal EPA communication].
[279] ACS, *American Chemical Society Annual Report 1996* (Washington, DC: ACS, 1997), 3.

Awards, established by President Clinton, which the Environmental Protection Agency and ACS helped design. The awards recognize and promote fundamental research and innovative green chemistry technologies, for example, chemical methods that prevent pollution and have broad applications in industry. Several of our members served on the panel that developed the [Presidential] Green Chemistry Challenge.

In this statement, the ACS explicitly recognized social concerns regarding the environment and the shift to pollution prevention policies. It was argued that both fundamental and applied chemical research was important in this respect and, consequently, the ACS served on the panel of the new awards. The ACS also reported another important reason to support the green chemistry initiative:[280]

> The American public is highly interested in environmental issues, and a common misconception is that chemists, chemistry, chemicals, and the chemical industry are all detriments to the environment. This perception presents a tremendous challenge to the Society to educate the public about what we do to maintain and improve our environment. In response to a request from EPA, ACS agreed to manage the selection panel for a new initiative, the Presidential Green Chemistry [Challenge] Award, that recognizes and promotes fundamental and innovative green chemistry technologies. ACS members served on the panel, and ACS President Ron Breslow gave the keynote address at the first ceremony, which was held at the National Academy of Sciences.

So clearly, ACS had a very strong interest in improving the public image of chemistry.

Since 1996, the awards were annually presented in addition to other green chemistry awards that were launched.[281] ACS's participation in the Presidential Green Chemistry Challenge Awards received political attention in 1997, as the ACS was rewarded for these efforts with a Hammer Award from the Vice President Gore's National Performance Review for these efforts.[282] This is interesting, because in the USA, federal government expenditures (% GDP) to non-defense related R&D, such as the chemical sciences, had been on the decline

---

[280] Ibid., 14. See: Ronald Breslow, "The Greening of Chemistry," *Chemical & Engineering News* 74 (1996): 72.

[281] "Green Chemistry Funding & Awards," ACS, accessed May 22, 2022, http://www.acs.org/content/acs/en/funding-and-awards/awards/gci.html. This website also reports other green chemistry awards.

[282] ACS, *A World of Information: American Chemical Society 1997 Annual Report* (Washington, DC: ACS, 1998), 3; Rosenblum, "Government Affairs," 64-74.

for decades.[283] Therefore, for ACS, the 1997 Hammer Award indicated that the term green chemistry was a powerful (rhetorical) tool that could have effect in the dialogue with the public, politicians included, in order to secure the legitimacy of the chemical sciences. A few years later the recognition of green chemistry by ACS received a new boost with the founding of the Green Chemistry Institute (GCI).

As demonstrated above a few scientists affiliated with EPA played major roles in organizing the pollution prevention symposia. One of them was Joseph Breen, who had graduated in chemistry at Fairfield University and received his PhD from Duke University. He was member of several ACS Divisions, e.g., DEC and the Division of Chemical Education. He served on ACS editorial boards and chaired the CEI Subcommittee Pollution Prevention. In the early 1990s, he was appointed as Chief of the Industrial Chemistry Branch (OPPT).[284] Breen received several EPA medals for his works. In the mid-1990s he retired from EPA.

But Breen was an ambitious person with a large network in the chemical arena. When the GCI was created in 1997, Breen became its first executive director. The institute was founded by stakeholders from academia, government, NGOs and industry.[285] GCI operated as an autonomous, independent, non-profit institute with no physical location. The primary objective of GCI was to promote green chemistry within the chemical sciences in the USA and worldwide, along the lines formulated by EPA policy officers such as Anastas and Breen.[286] In order to achieve this objective, Breen was responsible for the day-to-day

---

[283] ACS, *A World of Information: American Chemical Society 1997 Annual Report*, 13. See also: Susan Turner, "Joint Board-Council CEI Summary Report," *EnvirofACS* 44 (1996): 3; Anonymous, "Facts & Figures," 32-60; Hather et al., "The United States of America," 1-9. For concerns of the chemical industry in this respect, see: ACS, American Institute of Chemical Engineers, Chemical Manufacturers Association, Council for Chemical Research and Synthetic Organic Chemical Manufacturers Association, *Technology Vision 2020: the U.S. Chemical Industry* (Washington, DC: ACS, 1996), 16. The White House Office of Science and Technology supported this study in order to receive advice for the US government.

[284] Anonymous, "Election Issue," *EnvirofACS* 42 (1994): 12; Anonymous, "Election Issue," *EnvirofACS* 45 (1997): 10.

[285] Paul T. Anastas et al., "Promoting Green Chemistry Initiatives: Supported by a Rapidly Growing Infrastructure the Field Promises Innovative Solutions to Pressing Environmental Problems," *Environmental Science and Technology* 33 (1999): 116A-119A; Kira J.M. Matus, "The ACS Green Chemistry Institute®: a Case Study of Partnerships to Promote Sustainability in the Chemical Enterprise," in *Enhancing the Effectiveness of Sustainability Partnerships*, ed. Derek Vollmer (Washington, DC: The National Academies Press, 2009), 263-292, accessed May 22, 2022, www.nap.edu/html/12541/12541_casestudies_app.pdf; Roberts, "Creating Green Chemistry," 7-8; Paul T. Anastas, "Joe Breen – Heart and Soul of Green Chemistry," *Green Chemistry* 1 (1999): G87.

[286] Breen and Anastas were thus not practicing chemists. See also: Roberts, "Creating Green Chemistry," 8.

activities. He worked under the GCI Board, which included also Anastas and Denny Hjeresen (Los Alamos Laboratory). The first actions of GCI were focused on outreach, creating informal networks and the organization of symposia. The costs of these activities were covered by the initial funds of GCI: grants from industry and government. Unfortunately, Breen became seriously ill and passed away already in 1999.[287]

Hjeresen became the new executive director of GCI.[288] Under his leadership, GCI continued its efforts for the recognition of green chemistry. The institute did not do in-house chemical research but made use of external networks. GCI cooperated for instance with ACS and EPA, when the latter organization provided the ACS Division of Education with $ 240,000 in order to develop green chemistry education materials for students.[289]
In 2000, the GCI was absorbed by the ACS. In the Annual Report of the ACS it says:[290]

> Outside the United States, too, we are stronger than ever. This year, through formalized agreements, the prestigious Green Chemistry Institute became part of ACS operations in Washington, DC, and will promote the discovery and design of chemical products and processes that help save the environment.

and[291]

> The merger gives ACS a stronger presence in the global environmental arena and the institute the resources it needs to expand and enhance its programs. (…) GCI was a lead organizer of ChemRAWN XIV, which had a primary focus on green chemistry. Sponsored by the International Union of Pure and Applied Chemistry, ChemRAWN brought together some 200 scientists and policymakers from more than 30 countries to discuss the role of chemical

---

287 Minutes of the DEC Executive Committee Meeting, August 22, 1999, New Orleans, LA. Members at this meeting acknowledged the contributions of Breen. It was proposed to name an award after Breen. This became efficacious in 2000 with the Joseph Breen Memorial Fellowship. See also: ACS, "Green Chemistry Awards".
288 Anonymous, "Denny Hjeresen," *Green Chemistry* 5 (2003): G54-G55.
289 ACS, *A Combination of Many Different Elements: American Chemical Society Annual Report 1998* (Washington, DC: ACS, 1999), 12. The ACS also assisted in the organization of the "2nd Annual Green Chemistry and Engineering Conference: Global Perspectives". At the 218th ACS National Meeting a green chemistry symposium was organized by DEC and the ACS Division of Chemical Education. Anastas and Williamson lectured at this symposium. See: ACS, "218th American Chemical Society National Meeting & Exposition Program. New Orleans: August 22-26, 1999," *Chemical & Engineering News* 77 (1999): 59.
290 ACS, *American Chemical Society Annual Report 2000* (Washington, DC: ACS, 2001), 3.
291 ACS, *Our 125th Anniversary: Charting our Course for the 21st Century: American Chemical Society Annual Report 2001* (Washington, DC: ACS, 2002), 3 and 12.

research and science policy in advancing global environmental protection and sustainable development.

In contrast with for example DEC, GCI did not become a full technical division but continued to promote green chemistry as an institute. Phrases as ‘outside the United States, too, we are stronger than ever’, ‘stronger presence’ and ‘will promote’, as well as the relation with ‘policymakers’ and the environment demonstrate that green chemistry had become an important (rhetorical) instrument in the external communication of the ACS.

GCI received a start-up grant from ACS.[292] In addition to this financial aid, GCI tried to obtain funding from industries or governmental agencies, such as EPA and NSF.[293] Hjeresen left GCI in 2004 when he became Pollution Prevention and Sustainability Manager of the Los Alamos Laboratory.[294] His successor came from the White House Office of Science and Technology Policy (OSTP): Paul Anastas, one of the Godfathers of green chemistry in the US. After his years at EPA, Anastas had moved to the heart of USA’s science policy domain for some years, but in 2004 he became the fulltime GCI director. Probably, that hints at a close connection between GCI and OSTP, which could have been an additional reason for the ACS to support green chemistry.

As shown by its involvement in the green chemistry awards and the absorption of GCI, green chemistry was for ACS an opportunity to improve the social and financial ‘conditions’ for the chemical sciences.[295] GCI advocated, for instance, in cooperation with the ACS’s Office of Legislative and Government Affairs, for the Green Chemistry Research and Development Act of 2004 in order to obtain more R&D funding for the chemical sciences.[296] On the basis of green chemistry rhetoric’s, the chemical sciences got an improved stage at the

[292] This grant was mainly employed for expanding its staff. See: Matus, “The ACS Green Chemistry Institute®,” 263-292; Jennifer Young, interview, July 23, 2010. In 2004, Young started working for GCI and in 2009 she became manager of GCI. See also: ACS, *Vision: American Chemical Society Annual Report 2002* (Washington, DC: ACS, 2003), 12.
[293] Matus, “The ACS Green Chemistry Institute®,” 272-273; Young, interview, July 23, 2010.
[294] Anonymous, “Denny Hjeresen,” G54-G55.
[295] Hessels and Van Lente, “Practical Applications as a Source of Credibility,” 218.
[296] ACS, *Crossing Boundaries >> Crossing Disciplines: American Chemical Society 2004 Annual Report* (Washington, DC: ACS, 2005), 21. In its 2007 annual report, the ACS reported similar efforts. See also: ACS, *Improving People’s Lives through the Transforming Power of Chemistry: American Chemical Society Annual Report 2005* (Washington, DC: ACS, 2006), 27; Matus, “The ACS Green Chemistry Institute®,” 263-292; Young, interview, July 23, 2010.

interface of society, politics and science. President George Bush Jr. stated for instance in 2004:[297]

> I encourage today's leaders in science and industry to work together in the design, discovery and implementation of green chemistry principles. In addition, I salute the educational initiatives that help students learn about green chemistry and how it can be used to ensure our nation's economic and environmental prosperity.

But with respect to the recognition of green chemistry within ACS there were setbacks too. At the end of 2006, Anastas was appointed at Yale University, and left GCI.[298] His track record in green chemistry was strong. In 1996 he had proposed a green chemistry "philosophy", which subsequently culminated in 1998 with the 'twelve principles' of green chemistry that he published, jointly with John Warner (University of Massachusetts), in the very first green chemistry handbook.[299] In its 2005 annual report the ACS reported on the twelve principles of green chemistry:[300]

> The ACS Green Chemistry Institute (GCI) and its director, Paul T. Anastas, were named to the "Scientific American 50" – the magazine's prestigious annual list recognizing outstanding acts of leadership in science and technology. GCI, which became allied with the ACS in 2001, was saluted for its stewardship of the annual Presidential Green Chemistry Challenge Awards. Anastas was named for his work in developing the "Twelve Principles of Green Chemistry", key concepts that form the intellectual basis for the field.

The ACS presented the green chemistry 'philosophy' of Anastas and Warner as the 'intellectual basis for the field'.[301] It did not prove to be easy to find a successor to Anastas though; only in March 2008 Robert Peoples was appointed the new director.[302]

---

[297] ACS, *Crossing Boundaries >> Crossing Disciplines: American Chemical Society 2004 Annual Report*, 21. See also: Roberts, "Creating Green Chemistry," 138-140.
[298] Matus, "The ACS Green Chemistry Institute®," 263-292; Young, interview, July 23, 2010.
[299] Anastas and Williamson, "Green Chemistry," 1-17; Paul T. Anastas and John C. Warner, *Green Chemistry: Theory and Practice* (New York, NY: Oxford University Press, 1998), 29-55. See Chapter 5 for the epistemic components of the 12 principles. For the first joint publication of Warner and Anastas, see: Warner, Anastas and Anselme, "The Wittig Reaction," 346.
[300] ACS, *Improving People's Lives*, 27.
[301] Although, according to Young there were in 2010 still ACS members who were not aware of GCI, despite its appearance in ACS annual reports: Young, interview, July 23, 2010. See also Chapter 5..
[302] Matus, "The ACS Green Chemistry Institute®," 263-292; Young, interview, July 23, 2010.

In the meantime contours of the chemical sciences were changing. In 2005 the ACS reported that there was 'a growing number of chemists gravitating towards areas of shared interest among other scientific disciplines. ACS is staying abreast of these developments'.[303] To anticipate these changes, the ACS launched new journals, such as *ACS Chemical Biology*. But the field of green chemistry and its twelve principles, to which ACS referred in its annual report of 2005, was not recognized with a scientific ACS journal. *Green Chemistry Letters and Reviews* was launched in 2007 outside the ACS, under editorial leadership of Warner.[304]

2.7 Discussion: Green Chemistry and its Recognition

This chapter shows that both the negative image of chemistry and environmental concerns were drivers of great importance behind the emergence of green chemistry. The first initiatives to create green chemistry were rooted in the environmental chemistry division (DEC) of the ACS and in the environmental controlling agency EPA. These initiatives were supported by political decisions and by the contentious civic epistemology in the USA. All these factors were dominant for the (institutional) development of green chemistry and gave dominant non-laboratory scientists of EPA (such as Anastas) the opportunity to spread their views on green chemistry, as a I will further discuss below.

As argued above, in the 1960s, public concerns regarding the environment grew and were followed by political actions. These social concerns affected the operations of ACS over the subsequent decades. For instance, the membership of DEC grew quite steeply during the 1970s and 1980s (cf., Figure 2.3). Despite its active response to environmental issues, the ACS was plagued by a growing negative image of chemistry among the public. In the economic crisis of the early 1980s the Reagan Administration cut EPA's budget (see Figures 2.1 and 2.2). But after the environmental policies of the Reagan administration were heavily criticized, just before new elections, the Reagan Administration increased the EPA budget again and even introduced new environmental legislation that guaranteed more public transparency about environmental information. As a result, environmental concerns increasingly became a sociopolitical issue among ever larger audiences. Unintentionally, this

---

[303] ACS, *Improving People's Lives*, 13.

[304] Warner had published with Anastas the twelve principles of green chemistry in: Anastas and Warner, *Green Chemistry: Theory*, 29-55. See also: "Green Chemistry Letters and Reviews," Taylor & Francis Online, accessed May 22, 2022, http://www.tandfonline.com/toc/tgcl20/current.

created more legal polarization between industry, government and society when EPA's monitoring activities increased.

By 1990 it was recognized increasingly that the permanent legal fights between industry, government and society in USA's contentious civic epistemology were in need of a totally new approach.[305] Supported by worries about a bad economy (Figure 2.1), the legal mandate of EPA shifted from command-and-control objectives to pollution prevention policies, when the PPA was accepted by the US Congress.

Despite the fact that public attitudes changed slowly, the PPA was a turning point with respect to the emergence of green chemistry.[306] This act mandated EPA to support the development of pollution prevention and the chemical knowledge needed for that. That development was strategically shaped by mobilizing traditional networks within ACS though in particular DEC, as witnessed by the DfE activities such as the pollution prevention symposia mentioned above. This led to the first major green chemistry publication: *Green Chemistry: Designing.* Although there were some other technical divisions that co-sponsored pollution prevention symposia, this also shows that the ACS origins of green chemistry are principally within DEC due to DEC-members affiliated with EPA, such as Anastas.

Moreover, EPA attempted to put an end to the "triangle of distrust" that had contributed to a negative public image of chemistry. That distrust was politically undesirable as witnessed by *Reinventing*. Consequently, a strategic action was found necessary, also in the interest of ACS, to improve the image of chemistry and to decrease the distrust in USA's political culture. In 1995 the Presidential Green Chemistry Challenge Awards were announced, and green chemistry became soon a popular term within DEC and EPA. These awards and a new environmental act (PPA) show that the emergence of green chemistry was in part a result of political decisions.

In 2000, green chemistry received even stronger recognition within the US chemical community when the ACS absorbed GCI. This institute was under the leadership of dedicated individuals that stood at the basis of green chemistry in the USA, most notably Breen and Anastas. Thereby, the ACS had the aim to also improve the public image of chemistry. Applying the 'science-society contract' concept of Laurens Hessels and Harro van Lente, the ACS thus tried to change the 'identity' of chemistry in a "green" manner, in order to show the

---

[305] Jasanoff reports a contentious political culture for other domains too, which is characterized by a style of policy making that is legally driven and based on distrust. Jasanoff, *Designs on Nature*, 247-271.

[306] Roberts, "Creating Green Chemistry," 10.

relevance of chemistry to society.[307] Thereby the ACS also hoped to improve the funding 'conditions' for the chemical sciences. Seen from that perspective, I conclude that the term green chemistry thus gained institutional recognition within ACS for a number of reasons, with a rather strong symbolic and image dimension that was employed in dialogues with the public and politicians. Green chemistry was a powerful "rhetorical tool" for ACS to show society that chemistry was positively engaged with the environment.[308]

By the mid-2000s, after the green chemistry movement existed in the USA for ten years, the situation was ambivalent within the ACS. On the one hand, green chemistry received a significant recognition when it concerned improving the image of chemistry and the dialogue with the public, as demonstrated by the continuing ACS involvement in the Presidential Green Chemistry Challenge Awards program and GCI. On the other hand, the *scientific* recognition of green chemistry within ACS was a different matter. It seems to have lacked scientific status, as demonstrated by the long search for a new GCI director to succeed Anastas, and by the absence of a green chemistry based ACS journal.[309] That ambivalence in recognition demonstrates a vulnerability of green chemistry, which, as I conclude, is rooted in the way green chemistry was institutionalized within ACS. Whereas the technical division focused on the circulation of academic knowledge, GCI operations were directed to the interface of science, society and politics. Next to that, the ambivalence also related to the chemical foundations, scientific identity and internal scientific values of green chemistry, as will be discussed in Chapter 5.

---

[307] Hessels and Van Lente, "Practical Applications as a Source of Credibility," 218.
[308] Diana Crane, "Fashion in Science: Does it Exist?," *Social Problems* 16 (1969): 435.
[309] Roberts, "Creating Green Chemistry," 11 and 134. Garant and Young both confirmed a lack of scientific recognition of green chemistry by other chemists. Young, interview, July 23, 2010; Garant, interview, July 22, 2010.

Chapter 3: Recognition and Institutionalization of Green Chemistry in the UK

The previous chapter analyzed the rise of "green chemistry" in the USA by focusing on pollution prevention policies, on the role of the US Environmental Protection Agency (EPA) and on the American Chemical Society (ACS). A landmark in that development was the absorption in 2000 of the Green Chemistry Institute (GCI) by the ACS.

In view of a systematic comparison, this chapter follows the analytical framework developed in Chapter 2. Thus, public environmental concerns in the UK are briefly described by paying special attention to changing government policies on the environment and on pollution prevention. These policies are discussed against the background of UK's political culture and together the elements form the sociopolitical context in which chemists and the national chemical societies operated. In that light, I analyze the history and identity of environmental chemistry within the nationwide chemical societies in the UK. In 1972, an amalgamation took place of the Chemical Society (CS), the Society for Analytical Chemistry, the Royal Institute of Chemistry (RIC) and the Faraday Society. Eight years later the four societies fully merged, under the new name Royal Society of Chemistry (RSC).[310] Currently, the RSC has over 54,000 members and internationally the RSC has been recognized as one of the largest and most important chemical societies.[311]

As pointed out by David Whiffen and Donald Hey, the RSC has been both an independent learned society and a professional body.[312] Consequently, the historical analysis of environmental chemistry within the RSC provides the opportunity to mirror the sociopolitical demand for scientific knowledge regarding the environmental impact of chemicals in the UK from the 1970s onwards to 2000. Moreover, it sheds light on historical similarities and dissimilarities in comparison with the ACS Division of Environmental Chemistry (DEC) and with the emergence and development of green chemistry in the USA. The origins of the shift from environmental chemistry to green chemistry within the UK will be discussed. This shift culminated in the late 1990s in the establishment of a scientific journal and a green chemistry institute. The underlying mechanisms and forces influencing

---

[310] David H. Whiffen and Donald H. Hey, *The Royal Society of Chemistry: The First 150 Years* (London: RSC, 1991).
[311] "About the Royal Society of Chemistry," RSC, accessed May 22, 2022, http://www.rsc.org/about-us/.
[312] Whiffen and Hey, *The Royal Society of Chemistry*.

the institutionalization and recognition of green chemistry in the UK will be analyzed. I will also pay attention to environmental concerns within the public at large and to the changing public image of chemistry. The leading questions in this chapter therefore are: which were the dominant factors impacting the origins, growth and recognition of green chemistry in the UK? And why are these factors dominant?

3.1 Environmental Policy Tensions: Politics, the Economy and the Public

Similar to other Western countries, the British economy grew during more than two decades after the end of World War II, with rising standards of living.[313] Science-based industries were part and parcel of that growth.[314] But, there were also serious environmental drawbacks. For instance, the industrial and domestic burning of coal contributed to the so-called London Fog of 1952.[315] The Clean Air Act became a law in 1956, and in 1957 a national charity, the National Society for Clean Air, was launched.[316] Public environmental awareness thus grew, both nationally and internationally, on a wide range of topics.[317]

The UK government responded to these public voices. In 1965 the Natural Environment Research Council (NERC) was created, and on 11 December 1969, Prime Minister Harold Wilson (Labour) announced in the House of Commons a new advisory council: the Royal Commission on Environmental Pollution (RCEP), which was installed in

---

313 Tom Clark and Andrew Dilnot, "British Fiscal Policy since 1939," in *The Cambridge Economic History of Modern Britain: Structural Change and Growth, 1939-2000*, ed. Roderick Floud and Paul Johnson (Cambridge: Cambridge University Press, 2004), 384.

314 Martin Pugh, *State & Society: A Social and Political History of Britain 1870-1997* (London: Arnold Publishers, 1999), 254-296; Stephen J. Lee, *Aspects of British Political History 1914-1995* (London: Routledge, 1996), 179-228.

315 Nowadays, London fog is better known as particulate matter, which brings about diseases such as lung cancer.

316 *Clean Air Act 1956* (London: HMSO Publications, 1956). In The National Archives; available from http://www.legislation.gov.uk/ukpga/Eliz2/4-5/52/enacted, accessed May 22, 2022; S.J. Woodin, "Environmental Effects of Air Pollution in Britain," *Journal of Applied Ecology* 26 (1989): 749-761; Brian W. Clapp, *An Environmental History of Britain since the Industrial Revolution* (London: Longman Group UK Limited, 1994), 43-70; John Sheail, "'Burning Bings': a Study of Pollution Management in Mid-twentieth Century Britain," *Journal of Historical Geography* 31 (2005): 134-148.

317 Secretaries of State for Environment, Trade and Industry, Health, Education and Science, Scotland, Transport, Energy and Northern Ireland, the Minister of Agriculture, Fisheries and Food and the Secretaries of State for Employment and Wales, *This Common Inheritance: Britain's Environmental Strategy* (London: HMSO Publications, 1990), 32-34; Hannah Gay, "Before and After *Silent Spring*: from Chemical Pesticides to Biological Control and Integrated Pest Management-Britain, 1945-1980," *Ambix* 59 (2012): 88-108; Clapp, *An Environmental History of Britain*, 43-69; Susan Owens, "Experts and the Environment – the UK Royal Commission on Environmental Pollution 1970-2011," *Journal of Environmental Law* 24 (2012): 1-22.

1970.[318] This science-based 'committee of experts' had the following mandate to provide 'outside stimulus to government':[319]

> (…) to advise […] on matters, both national and international, concerning the pollution of the environment; on the adequacy of research in this field; and the future possibilities of danger to the environment.

Wilson also announced the establishment of the Central Scientific Unit on Pollution that 'intended to coordinate action within government', which was installed in 1970 too.[320]

The most important environmental law during those years was enacted in 1974: the Control of Pollution Act (CPA), which provided new controls regarding noise, water and air pollution, as well as waste disposal.[321] This law led to an increasing demand for new scientific knowledge by several stakeholders ranging from 'water authorities' to the 'Secretary of State'.[322] Based on CPA, also many agencies, including local authorities, received the power and authority to control a specific segment of the environment (such as air, water, soil). Therefore, according to the political scientists James Connelly and Graham Smith, 'the organisation and structure of pollution control in Britain has traditionally been fragmented'.[323] This also led to 'negotiated consents' between public officials and representatives from industry, which contributed to a governmental pollution control in the UK that was:[324]

---

318 Gay, "Before and After *Silent Spring*," 95; "Vision and Purpose," NERC, accessed May 22, 2022, http://www.nerc.ac.uk/about/whatwedo/vision/; Owens, "Experts and the Environment," 2; Derek Osborn, "Some Reflections on UK Environment Policy, 1970-1995," *Journal of Environmental Law* 9 (1997): 3-22. From 1970 until 1995, Osborn was a civil servant at the Department of the Environment (DoE).

319 Owens, "Experts and the Environment," 1-2. Owens is a former member of the RCEP.

320 Ibid.; Osborn, "Some Reflections on UK Environment Policy," 4. The Central Scientific Unit on Pollution was soon absorbed within the DoE.

321 *Control of Pollution Act 1974* (London: HMSO Publications, 1974). In The National Archives; available from https://www.legislation.gov.uk/ukpga/1974/40, accessed May 22, 2022.

322 Ibid., 98-103. See the section "Research and publicity" of this law.

323 James Connelly and Graham Smith, *Politics and the Environment: From Theory to Practice* (London: Routledge Taylor & Francis Group, 2003), 300. See also: *Control of Pollution Act 1974*, 102.

324 Connelly and Smith, *Politics and the Environment*, 300. See also: T. O'Riordan, "Industrial Pollution Control in the UK," *The Science of the Total Environment* 129 (1993): 39-53. O'Riordan argues that the regulatory regime was, more or less, 'incestuous'. See for example: R.C. Paul, "An Industrialist's View," in *Industry and the Environment in Perspective*, ed. R.E. Hester (London: RSC, 1983), 12-29. As an industrialist, Paul preferred consultation, instead of confrontation, between industries and local authorities. In the 1981-1983 period, Hester (York University) was Chairman of the RSC Environmental Chemistry Group. This network will be discussed in the next section.

> (…) based on a philosophy of "partnership" between the regulator and industry [and therefore] very few prosecutions occurred. The traditional politics of pollution control in many ways mirrored the rest of British political culture, with a close and often unspecified relationship between government and its agencies and the industrial sector.

After decades of prosperity the economy became a major political concern in Europe and the USA from the mid-1970s onwards due to very high oil prices and competition by low-wages countries.[325] This led to unemployment rates in the UK that were significantly higher than, for example, the USA and France.[326] In 1981, the UK officially was in an economic recession.[327] Economic problems also led to polarization in UK's political culture as illustrated by the elections of 1979, which were won by the Conservatives. Margaret Thatcher became the new Prime Minister after a campaign that was fueled with anti-socialist and anti-Keynesian rhetoric's. She argued, for instance, 'that socialism has failed to fulfil its promises'.[328]

These political views also impacted UK's environmental policies, which started to deviate increasingly from other countries. Liberal governmental policies, also with respect to the environment, now came to the fore.[329] Thatcher wanted to restore market forces and also a reduction of state intervention. A number of state-owned companies and assets were privatized. In this economic, political and sociocultural spirit of time, the environmental momentum of the late 1960s and early 1970s declined, at least in governmental policies. The existing environmental legislation reinforced 'UK Conservative government's ideological opposition' to 'the development of a comprehensive environmental policy, which it saw as an impediment to economic development', according to James Connelly and Graham Smith.[330] Also economically and socio-culturally the environmental momentum declined, because the public was more worried about basic needs instead of the environment. According to Derek Osborn, a former Director General of Environment Protection and Chairman of the European Environment Agency, this weak public environmental concern was more obvious in the UK

---

[325] Pugh, *State & Society*, 329-344; Lee, *Aspects of British Political History*, 238.
[326] Mary O'Mahony, "Employment, Education and Human Capital," in *The Cambridge Economic History of Modern Britain*, Floud and Johnson, 116. According to Pugh, registered unemployment increased from 542,000 in 1974 to 3,149,000 in 1985: Pugh, *State & Society*, 347.
[327] Clark and Dilnot, "British Fiscal Policy," 391-392.
[328] Quote of Thatcher in 1977, which is taken from: Pugh, *State & Society*, 344.
[329] Lee, *Aspects of British Political History*, 237-238; Clark and Dilnot, "British Fiscal Policy," 391-392.
[330] Connelly and Smith, *Politics and the Environment*, 295.

than in other Northern European countries.[331] In the early 1980s the proportion of the UK population that was member of an environmental interest group was approximately 5.0 %, which was significantly lower than for instance in the Netherlands.[332]

The UK government avoided the inclusion of environmental pressure groups within the process of policy making. This feature of British environmental policy did not change in the late 1980s, although public concerns regarding the environment started to grow again, as witnessed by increasing membership numbers of for example Friends of the Earth, founded in the UK.[333] Scientific knowledge regarding the impact of chemicals on the environment grew as well, and this reinforced feelings of public anxiety in some cases, for instance as a result of environmental disasters.[334] By contrast, the Thatcher Administration, according to Osborn, emphasized scientific uncertainties. It ignored critical voices of the public and insisted on its liberal doctrine and policies.[335] Consequently, until the late 1980s Britain had loose environmental quality standards, a fragmented organization of pollution control, and a government that was not very transparent with respect to its collaboration with industry on environmental issues. In comparison with 1980, UK's SO_2 emissions had dropped by 22.8 % in 1990, whereas for example the Netherlands witnessed a 55.6 % decrease.[336] According to some, the UK was therefore known as 'the dirty man of Europe'.[337] During the 1980s the situation changed though, under the influence of the environmental policies within the

---

[331] Osborn, "Some Reflections on UK Environment Policy," 4; Derek Osborn, preface to European Environment Agency, *Annual Report 1999* (Copenhagen: European Environment Agency, 2000), 5-6.
[332] See for an overview: Russell J. Dalton, "The Greening of the Globe? Cross-national Levels of Environmental Group Membership," *Environmental Politics* 14 (2005): 441-459.
[333] Neil Carter, *The Politics of the Environment: Ideas, Activism, Policy* (Cambridge: Cambridge University Press, 2010), 146; "Our History," Friends of the Earth UK, accessed May 22, 2022, https://friendsoftheearth.uk/who-we-are/our-history. Friends of the Earth was founded in 1971. See also: Osborn, "Some Reflections on UK Environment Policy," 7; Clapp, *An Environmental History of Britain*, 55-60; Owens, "Experts and the Environment," 10; Geoff Dickes, "A Brief History of the Environmental Chemistry Group," *Environmental Chemistry Group Newsletter* 1 (1995): 4-5; Woodin, "Environmental Effects," 749-761; P.A. Cawse, "Inorganic Particulate Matter in the Atmosphere," in *Specialist Periodical Reports: Environmental Chemistry Volume 2*, ed. H.J.M. Bowen (London: RSC, 1982), 69.
[334] See for example: David Price, "Sink or Swim?: Ozone Depletion, Friends of the Earth, Consumers' Association, and European Campaign Against Cancer," *Energy Policy* 17 (1989): 295-296.
[335] Osborn, "Some Reflections on UK Environment Policy," 5.
[336] United Nations Economic Commission for Europe, *Updated Handbook for the 1979 Convention on Long-range Transboundary Air Pollution and its Protocols* (Geneva: United Nations, 2015), 110.
[337] Timothy O'Riordan, "The Politics of Environmental Regulation in Great Britain," *Environment* 30 (1988): 4-9. By then, O'Riordan was professor of environmental sciences and served on a number of environmental advisory committees in the UK.

European Economic Community (EEC), and the publication of *Our Common Future* in 1987.[338]

Starting in the mid-1980s, countries such as Germany and Denmark increasingly argued for a 'Community-wide pollution control process' within the EEC in order to improve the quality of the environment.[339] At first, the Thatcher Administration, which insisted on British sovereignty and opposed to stringent environmental policies, lacked a willingness to support these EEC policy initiatives. In the late 1980s that political attitude shifted though. For example, an advice by the RCEP for an integrated pollution control, dating already from 1976, finally received political recognition in 1987.[340] In the same year a fundamental institutional change took place with the establishment of Her Majesty's Inspectorate of Pollution (HMIP). HMIP was a response to the growing environmental concerns of the British society and increasing pressures of the EEC.[341] The UK government was also influenced by the well-known Brundtland Report of 1987. Stewardship was now emphasized as the ethical imperative for future environmental policies, which was equivalent to the norm of sustainable development, that is to look after your planet now, and for future generations.[342]

That change in government attitudes is illustrated by the following quote from a public address given by Margaret Thatcher in 1988 to the Royal Society:[343]

> For generations, we have assumed that the efforts of mankind would leave the fundamental equilibrium of the world's systems and atmosphere stable. But it is possible that with all these enormous changes (population, agricultural, use of fossil fuels) concentrated into such a short

---

338 United Nations World Commission on Environment and Development, *Our Common Future* (Oxford: Oxford University Press, 1987). This report is also known as the "Brundtland Report".

339 Connelly and Smith, *Politics and the Environment*, 303. See also: Council of the European Communities and Commission of the European Communities, *Treaty on European Union* (Luxembourg: Office for Official Publications of the European Communities, 1992), 12; Sonja A. Boehmer-Cristiansen, "Emerging International Principles of Environmental Protection and Their Impact on Britain," *Environmentalist* 10 (1990): 95-113; Marie Soveroski, "EC Enlargement and the Development of European Environmental Policy: Parallel Histories, Divergent Paths?," *Review of European Community and International Environmental Law* 13 (2004): 127-134. Soveroski focuses on the European Community (EC).

340 Owens, "Experts and the Environment," 11-14; Connelly and Smith, *Politics and the Environment*, 302-303. See for the increase in staff numbers: HMIP, *Annual Report 1993-94* (London: HMSO Publications, 1994), 117.

341 Owens, "Experts and the Environment," 11.

342 Secretaries of State for Environment et al., *This Common Inheritance*, 10. See also: United Nations World Commission on Environment and Development, *Our Common Future*. For a detailed overview about the influence of this report on UK's environmental policy, see: Connelly and Smith, *Politics and the Environment*, 309-310.

343 Margaret Thatcher, "Speech to the Royal Society: September 27, 1988," Margaret Thatcher Foundation, accessed May 22, 2022, http://www.margaretthatcher.org/document/107346.

period of time, we have unwittingly begun a massive experiment with the system of this planet itself.

The leading science journal *Nature* wondered therefore in 1988: 'Has Mrs. Thatcher turned green?'.[344] In 1989 the European Parliament elections resulted in an electoral success of the Green Party.[345] This political wake-up call also contributed to a shift in UK's environmental policy.[346]

Although Thatcher's turn was probably more electorally motivated than ideologically, her administration presented in 1990 the very first UK White Paper on the environment: *This Common Inheritance: Britain's Environmental Strategy*.[347] The Thatcher Administration argued in this vision document with respect to the style of policy making and public accountability in case of pollution matters:[348]

> The Government has made access to environmental information a key part of Britain's new comprehensive system of pollution control, and it warmly supported the European Community Directive on this subject.

This statement reflected a partial farewell to UK's communitarian civic epistemology of the 1970s and 1980s in which environmental control was often determined in a relational and consultative way between industry and state (in order to secure UK's industrial competitiveness).[349] That earlier policy style had resulted in a significant lack of transparency and public participation in relation to the environment and its scientific study. But with improving the 'access to environmental information' this began to change. Until then the influence of the state, with its insulated governmental committees, had been huge as compared to the input of public voices, and the state had been rather immune to geopolitical pressures.[350] But, owing to the growing environmental concerns of the public and EEC

344 Anonymous, "Has Mrs Thatcher turned green?," *Nature* 335 (1988): 479.
345 Connelly and Smith, *Politics and the Environment*, 87-88.
346 Ibid.
347 Secretaries of State for Environment et al., *This Common Inheritance*. According to the UK's National Federation of Self Employed and Small Businesses this report was a 'government public relations exercise'. See heretofore: "Green Is Good but it Will Cost Firms Money," *Newcastle Journal*, October 25, 1990, 13.
348 Secretaries of State for Environment et al., *This Common Inheritance*, 12.
349 For an overview concerning UK's communitarian civic epistemology, see: Sheila Jasanoff, *Designs on Nature: Science and Democracy in Europe and the United States* (Princeton, NJ: Princeton University Press, 2005).
350 Ibid., 258-271.

influences, the environment became more and more a political issue by the end of the 1980s. As environmental policies had to be based on 'fact not fantasy', this also provided new opportunities for chemists to investigate environmental issues and to cope with scientific uncertainties.[351]

3.2 The Emergence of Environmental Chemistry and the RSC Environmental Chemistry Group

The establishment of the NERC in 1965 led to new funding that became available for scientists with environmental interests. This impacted the chemical sciences too.[352] For instance the 1971 Joint Annual Meeting of CS and RIC was almost entirely focused on the symposium "Chemistry and the Environment".[353] The RIC formulated technical advice to the UK government on legislative matters, which was based on its chemical knowledge of the environment.[354] According to George Porter, President of the CS, the public expected the chemist 'to explain and interpret his work to the non-scientist and to become involved in the problems of protection of the environment'.[355] So, in the early 1970s chemists were increasingly concerned about the environment and argued that proper scientific attention should be given to that.

In 1966, the Bylaws of the CS were modified leading to the formal establishment of "groups" within its divisions.[356] In 1972, the year of amalgamation, groups were formed within the newly created CS Industrial Division. In this year, some members of this scientific network met to discuss the major governmental responses to public environmental concerns, e.g., the CPA.[357] This led to the establishment of the Environment Group.[358] In the years

---

[351] Secretaries of State for Environment et al., *This Common Inheritance*, 10.
[352] Gay, "Before and After *Silent Spring*," 95.
[353] This symposium gained a lot of 'press interest' too. See: RIC, *Royal Institute of Chemistry: Report of the Council for the Year Ended 30 September of 1971* (London: RIC, 1971), 10; CS, *The Chemical Society: Annual Report of Council and Accounts 1970-1971* (London: CS, 1971), 7.
[354] Whiffen and Hey, *The Royal Society of Chemistry*, 219. Probably, in later years this task was fulfilled by the RSC Environment, Health and Safety Committee (EHSC). See: RSC, *The Royal Society of Chemistry: Annual Report 1986* (London: RSC, 1987), 4 and 14; RSC, *The Royal Society of Chemistry: 1988 Annual Report* (London: RSC, 1989), 18.
[355] George Porter, "The Amalgamated Societies," *Nature* 234 (1971): 511-512.
[356] Whiffen and Hey, *The Royal Society of Chemistry*, 25. Some groups existed informally prior to 1966.
[357] Dickes, "A Brief History," 4.
[358] Ibid. The 'first committee meeting' took place in 1973. For further information, see also: CS, *The Chemical Society: Annual Report of Council and Accounts 1970-1971*, 8; Whiffen and Hey, *The Royal*

following 1972, the impact of that group within the CS, and its successor the RSC, increased, as will be discussed below.

With the founding of the Environment Group, environmental chemistry increasingly received attention during the early 1970s. The objectives of the Environment Group were:[359]

> To promote understanding of the effects, both advantageous and disadvantageous, of chemicals from industry on the environment; to identify areas where solutions to existing problems need to be found; and to promote wider knowledge of the benefits accruing from chemicals on the environment.

The CS Industrial Division therefore often organized symposia that concerned 'environmental problems and legislation'.[360] The purposes of the Environment Group were thus basically a call for knowledge in order to understand the environmental impact of chemical industries. So from the early 1970s onwards, research activities of environmental chemists started to pay attention to these impacts, as witnessed by the *Specialist Periodical Reports: Environmental Chemistry* (*SPR*).[361] In 1975, the first volume of this series, on the organic aspects of environmental chemistry, was published under editorial leadership of Geoffrey Eglinton (University of Bristol). He argued that in research on environmental chemistry the 'emphasis often lies on those environments under stress in the industrialised areas of the world', i.e., those areas that were subjected to public scrutiny.[362] That scientific research was now named 'environmental chemistry' and defined by Eglinton:[363]

> (….) as the assessment of the distribution and interaction of elements and compounds in the environment, their modes of transport and their effect on biological and other systems.

Society of Chemistry, 206-209. This group did not informally exist prior to its establishing within CS, nor did any other environmentally engaged group.

[359] Dickes, "A Brief History," 4. Dickes is a former Chairman of this group.

[360] CS, *The Chemical Society: Annual Report of Council and Accounts 1973-1974* (London: CS, 1974), 9. See also: CS, *The Chemical Society: Annual Report of Council and Accounts 1970-1971*, 7; CS, *The Chemical Society: Annual Report of Council and Accounts 1974-1975* (London: CS, 1975), 6-9; Dickes, "A Brief History," 4; Whiffen and Hey, *The Royal Society of Chemistry*, 206-207.

[361] Since 1968, the *SPR* have been published by the CS and, subsequently, the RSC. The *SPRs* concern different topics of study ranging from photochemistry to catalysis. See: Whiffen and Hey, *The Royal Society of Chemistry*, 140-142.

[362] G. Eglinton, preface to *Specialist Periodical Reports: Environmental Chemistry Volume 1*, ed. G. Eglinton (London: CS, 1975), IV. See also: James Maxwell, "Professor Geoffrey Eglinton FRS, 1927-2016," University of Bristol, last modified March 24, 2016, www.bristol.ac.uk/news/2016/march/geoff-eglinton.html.

[363] Eglinton, preface to *Specialist Periodical Reports: Environmental Chemistry Volume 1*, III.

This definition characterizes the scientific knowledge published in *Environmental Chemistry Volume 1*. For instance, in this report the research done on the pathways of organic compounds, which were released in aquatic environments due to for example oil spills, was reviewed, as well as the bioaccumulation of polychlorinated biphenyls (PCBs).[364]

In 1982, *Environmental Chemistry Volume 2* was published under editorial leadership of Humphry Bowen (University of Reading).[365] This report concerned the inorganic aspects of environmental chemistry, hereby covering different topics ranging from the atmosphere to human diets.[366] Bowen argued that environmental chemistry should not follow the traditional lines of organic, inorganic and physical chemistry.[367] In 1984, *Environmental Chemistry Volume 3* was therefore published by ignoring these thematic lines leading to a wide variety of environmental research topics.[368] For instance, research done on SO_2 and heavy metals in suspended particulates released by the combustion of fossil fuels was discussed in relation to public health risks.[369] Clearly, a significant part of the research activities was related to anthropogenic actions ranging from chemical industries to agricultural practices.[370] In addition, some environmental chemists employed perspectives taken from geology, meteorology and geochemistry.[371] This type of knowledge provided a long-term perspective on contemporary anthropogenic influences on the environment. But, it also illustrates the

---

364 R.J. Morris and F. Culkin, "Environmental Organic Chemistry of Oceans, Fjords, and Anoxic Basins," in *Environmental Chemistry Volume 1*, Eglinton, 81-108; J.W. Farrington and P.A. Meyer, "Hydrocarbons in the Marine Environment," in *Environmental Chemistry Volume 1*, Eglinton, 109-136; P.A. Cranwell, "Environmental Organic Chemistry of Rivers and Lakes, Both Water and Sediment," in *Environmental Chemistry Volume 1*, Eglinton, 22-54; M.M. Rhead, "The Fate of DDT and PCBs in the Marine Environment," in *Environmental Chemistry Volume 1*, Eglinton, 137-159; J.E. Allebone, R.J. Hamilton and B. Ravenscroft, "Environmental Organic Chemistry of 2,4-Dichloro-phenoxyacetic Acid," in *Environmental Chemistry Volume 1*, Eglinton, 160-190. In addition, the CS explained the uses and hazards of herbicides at BBC radio, see: CS, *The Chemical Society: Annual Report of Council and Accounts 1979* (London: CS, 1980), 4.

365 Bowen passed away in 2001. See: J. Michael Moffatt, "Humphry Bowen: Practical Botanist and Chemist Exploring the Natural World," *The Guardian*, last modified October 17, 2001, https://www.theguardian.com/news/2001/oct/17/guardianobituaries.highereducation.

366 H.J.M. Bowen, preface to *Environmental Chemistry Volume 2*, Bowen.

367 Ibid.

368 H.J.M. Bowen, ed., *Specialist Periodical Reports: Environmental Chemistry Volume 3* (London: RSC, 1984).

369 Cawse, "Inorganic Particulate Matter," 1-69.

370 See also: I. Colbeck and R.M. Harrison, "Tropospheric Ozone," in *Environmental Chemistry Volume 3*, Bowen, 1-48. Roy Harrison was a former Chairman of the Environment Group. See: Dickes, "A Brief History," 4; "Professor Roy Harrison," University of Birmingham, accessed May 22, 2022, https://www.birmingham.ac.uk/staff/profiles/gees/harrison-roy.aspx. See: S.J. Blunden, L.A. Hobbs and P.J. Smith, "The Environmental Chemistry of Organotin Compounds," in *Environmental Chemistry Volume 3*, Bowen, 49-77.

371 A.M. Ure and M.L. Berrow, "The Elemental Constituents of Soils," in *Environmental Chemistry Volume 2*, Bowen, 94-204.

complexity of carrying out environmental chemistry research, which typically demanded for numerous data obtained by improved analytical chemistry techniques.

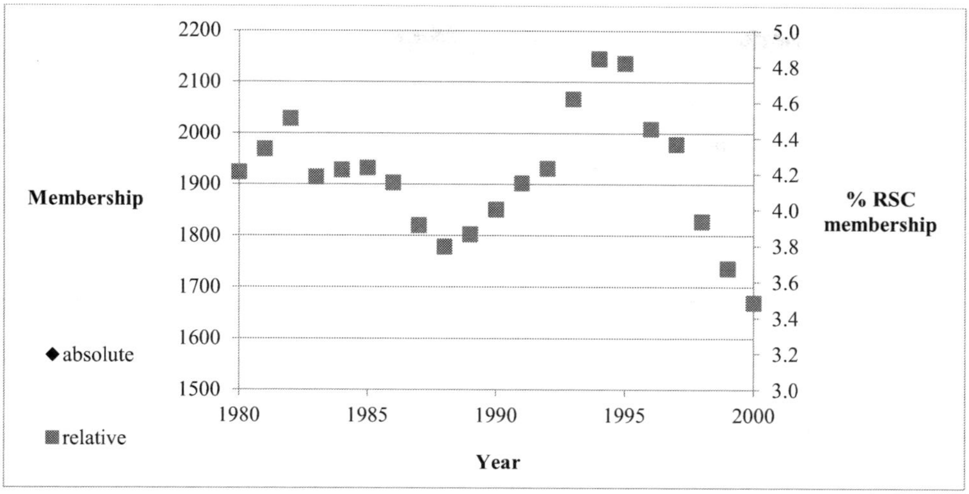

Figure 3.1: RSC Environmental chemistry membership for the 1980-2000 period.[372]

The impact of the Environment Group within the RSC was significant. The Environment Group was for instance consulted by the RSC leadership in matters of legislations concerning health, safety and the environment.[373] Moreover, Figure 3.1 shows that, in 1987, the Environment Group had 1591 members, i.e., 3.9 % of the total RSC membership. With the increased environmental interest described in the political arena starting in the late 1980s, the membership figures of the environmental chemistry group in the UK grew. This led in 1994 to the highest membership level for the 1980-2000 period, with 2168 members, i.e., 4.8 % of the total RSC membership.

A reorganization took place in the same year. In 1994 the Environment Group became formally independent from the RSC's Industrial Division: the RSC Environmental Chemistry Group (ECG) was founded. Environmental chemistry was now regarded as a separate subfield

[372] Data obtained from the RSC annual reports (1980-1994) and the RSC annual records (1995-2000). Data from CS, the predecessor of RSC, are not available.

[373] RSC, *The Royal Society of Chemistry: 1987 Annual Report* (London: RSC, 1988), 17. See also: Whiffen and Hey, *The Royal Society of Chemistry*, 218-220.

of the chemical sciences and a constitutional change within the RSC underlined that new identity:[374]

> From its inception in the early 1970s until 1994, the Environment Group operated as part of the RSC's Industrial Division. During this period environmental chemistry has grown and matured into a scientific discipline in its own right, and many of our 2000+ members are active in developing the chemistry of environmentally-related subjects. As a reflection of this change of emphasis, in 1993 the Environment Group Committee sought and received the members' approval to change the Group's name to the Environmental Chemistry Group and to switch our affiliation to the Society's Scientific Affairs Board.

ECG became thus independent under the RSC Scientific Affairs Board, although there was still cooperation with the RSC Industrial Division. Under the leadership of John Holder (University of Central Lancashire), fifteen environmental chemists, from industry, academia and regulatory work, became the quartermasters of this new group. The ECG was one of the largest scientific networks within RSC in the early 1990s.[375]

In a nutshell, the origins of environmental chemistry in the UK were rooted in the applied research interest of industrial chemists, with their end-of-pipe solutions to pollution caused by chemical plants. Triggered by the growing public environmental concerns and environmental legislations, this led in 1972 to an institutionalization in the form of the Environmental Group within the CS Industrial Division. Gradually, the research agenda became more influenced by fundamental research interests as witnessed by for instance the geochemical approaches and by an increasing awareness of complexity. Environmental pathways and impact of chemicals are not solely determined by anthropogenic influences, but by many other (chemical) interactions as well. Furthermore, with knowledge and techniques acquiring from toxicology, combined with for instance meteorology and analytical chemistry, environmental chemistry developed during the 1970s and 1980s into a scientific subfield that was spurred by interdisciplinary insights, and stayed like that the next decades. It was united by a common research object, the environment at large, and not by a specific scientific approach. As a result, the identity of environmental chemistry was also rather vague with respect to its approaches and boundaries within the chemical sciences and environmental

---

374 Anonymous, "The Environmental Chemistry Subject Group of the Royal Society of Chemistry," *Environmental Chemistry Group Newsletter* 1 (1995): 1.
375 John V. Holder, "Chairman's Report on Activities During 1994," *Environmental Chemistry Group Newsletter* 1 (1995): 2.

chemists were typically not engaged in the development of new chemical synthetic pathways.[376] The rise of environmental chemistry in the UK strongly related to the demand for environmental knowledge by industry, the public and the government. It was institutionalized for that reason in 1972 and became as the ECG independent under the RSC Scientific Affairs Board in 1994, i.e., shortly after the first UK White Paper on the environment and right in a period of a membership peak (cf., Figure 3.1).

3.3 Pollution Prevention and Technology Foresight: Setting the Stage for Green Chemistry

From the 1970s onwards, the CS and the RSC have been influenced by environmental concerns and policies, as witnessed by the rise of the environmental chemistry. Environmental chemists have often been dealing with waste that is released in the environment in order to identify the impact and pathways of pollutants. However, from an environmental point of view, prevention is better than cure. As pointed out in Chapter 2, at least 35 countries, the UK included, realized this by signing the science-based multilateral *1979 Convention on Long-Range Transboundary Air Pollution* (*LRTAP*), which entered into force in 1983.[377] *LRTAP* laid down policy principles, such as pollution prevention. This resulted in policy standards and targets that were formulated in binding protocols, which followed a consensus-based decision making. But, according to the political scientist Pamela Chasek and her colleagues, the UK often 'refused to join particular protocols or demanded that they be watered down'.[378] Indeed, until the late 1980s the UK politically insisted on its sovereignty and liberal doctrine regarding environmental policies. In this section I discuss that this approach changed at the

---

[376] In 1994, Manahan argued the same with respect to the identity of environmental chemistry: Stanley E. Manahan, *Environmental Chemistry* (London: Lewis Publishers, 1994), 3-4. See also: Bernadette Bensaude-Vincent and Jonathan Simon, *Chemistry: The Impure Science* (London: Imperial College Press, 2008), 6. Environmental chemists fit the general picture of chemistry as an "operational realism" discipline poorly, as these chemists are hardly involved in the design of new substances.
[377] Pamela S. Chasek, David L. Downie and Janet W. Brown, *Global Environmental Politics* (Philadelphia, PA: Westview Press, 2010), 121-129; Willemijn Tuinstra, Leen Hordijk and Carolien Kroeze, "Moving Boundaries in Transboundary Air Pollution Co-production of Science and Policy under the Convention on Long Range Transboundary Air Pollution," *Global Environmental Change* 16 (2006): 349-363; Adam Byrne, "The 1979 Convention on Long-Range Transboundary Air Pollution: Assessing its Effectiveness as a Multilateral Environmental Regime after 35 Years," *Transnational Environmental Law* 4 (2014): 37-67; United Nations, *Convention on Long-Range Transboundary Air Pollution* (Geneva: United Nations, 1979). In United Nations Treaty Collection; available from https://treaties.un.org/Pages/ViewDetails.aspx?src=TREATY&mtdsg_no=XXVII-1&chapter=27&clang=_en, accessed May 22, 2022.
[378] Chasek, Downie and Brown, *Global Environmental*, 128.

end of the 1980s, with a stronger emphasis on pollution prevention policies. These changing policies impacted the operations of the RSC, and they also contributed to the start of green chemistry in the UK.

The UK government already introduced pollution prevention initiatives prior to the publishing of *This Common Inheritance* (1990). For example, in 1988 the Department of the Environment (DoE) launched the Environmental Protection Technology Scheme (EPTS).[379] This was a partnership between the UK government and some industries that cooperated on voluntary grounds to promote pollution prevention by funding research projects. Initiatives such as the EPTS received some recognition, because prevention was environmentally 'better and cheaper than cure'.[380] The Thatcher Administration thus searched for environmental policies that were also good for the economy.

A major legal reform took place with the establishment of the Environmental Protection Act 1990 (EPrA), which was the outcome of political pressures by the EEC. EPrA was the legislative response of the Thatcher Administration to several environmental directives from the EEC since the mid-1980s, leading to for instance re-enacting provisions of older environmental laws.[381] In contrast with the Pollution Prevention Act of 1990 in the USA, a major objective of EPrA was to consider pollution processes 'as a whole' in order to determine their impact 'on the environment as a whole'.[382] This was principally an integrated approach towards the environment, in contrast to previous legal practices in the UK.[383] Considering different "Publicity" sections of the new law, another move away from traditional UK features was the increasing participation of the public in environmental policy making. For instance, the public had the right to comment on proposed licenses.[384] These objectives altogether had to be accomplished by employing the philosophy of Integrated

---

379 Secretaries of State for Environment et al., *This Common Inheritance*, 187.
380 Ibid., 11.
381 Connelly and Smith, *Politics and the Environment*, 264-265 and 298-303; O'Riordan, "Industrial Pollution Control," 41-44. Directives are legally binding member states in order to accomplish with E(E)C policy goals, but leave these states free in the way they enact with these goals. See also: *Environmental Protection Act 1990* (London: HMSO Publications, 1990). In The National Archives; available from www.legislation.gov.uk/ukpga/1990/43/pdfs/ukpga_19900043_en.pdf, accessed May 22, 2022.
382 Connelly and Smith, *Politics and the Environment*, 304.
383 Ibid., 304; "Pollution Prevention Act of 1990," in *Compilation of Selected Acts within the Jurisdiction of the Committee on Commerce: Environmental Law*, Committee on Commerce U.S. House of Representatives (Washington, DC: US Government Printing Office, 1995), 615-623.
384 Connelly and Smith, *Politics and the Environment*, 303.

Pollution Control (IPC), which was based on pollution prevention, another major purpose of EPrA.[385]

In 1991, HMIP operationalized IPC.[386] Or, to use the words of HMIP representative Robert Gemmill:[387]

> IPC provides HMIP with a combination of powers over processes and their operation as well as over releases to the environment. This means that HMIP is not wholly dependent on information derived from measurement of releases. Prevention rather than cure (emphasis on cleaner or low waste technologies rather than on end-of-pipe technology; on eliminating or minimizing the production of waste at source rather than on treatment of the wastes produced) is at the heart of HMIP's approach to its regulatory responsibilities. Nevertheless, monitoring remains a key element not only of regulatory control, but also of providing information to the public in the new climate of open regulation.

Clearly, in the early 1990s, UK's environmental policies shifted towards pollution prevention hereby following an integrated approach. This trend continued in the mid-1990s, as witnessed by the establishment of the UK's Environment Agency in 1996, which was a merger of several controlling agencies (with HMIP included).[388]

The focus on pollution prevention was politically emphasized in *This Common Inheritance*:[389]

> The Government's first priority is to reduce waste at source to a minimum. It intends to achieve this by imposing tough standards on industry through Integrated Pollution Control (IPC) and by promoting clean technologies.

---

[385] Ibid., 144.
[386] Ibid., 303.
[387] R.J. Gemmill, "The Approach of Her Majesty's Inspectorate of Pollution to Regulatory Monitoring," *Environmental Chemistry Group Newsletter* 3 (1996): 3-4. Gemmill was also ECG's Secretary. See: Anonymous, "The Environmental Chemistry Subject Group," 1. See also: Neil Emmott and Nigel Haigh, "Integrated Pollution Prevention and Control: UK and EC Approaches and Possible Next Steps," *Journal of Environmental Law* 8 (1996): 301-311.
[388] Connelly and Smith, *Politics and the Environment*, 305. With more than 10,000 staff, the EA became one of the biggest environmental agencies in the world. The EA was an outcome of the Environment Act 1995. See heretofore: *Environment Act 1995* (London: HMSO Publications, 1995). In The National Archives; available from http://www.legislation.gov.uk/ukpga/1995/25/section/4, accessed May 22, 2022.
[389] Secretaries of State for Environment et al., *This Common Inheritance*, 186.

This call for 'promoting clean technologies' and the aiming for pollution prevention, shows that another type of chemical knowledge was increasingly demanded, which impacted UK's science policy in the 1990s. In line with the UK's political culture, it was believed that 'commercial forces are the most effective spur for technological progress' in order to 'achieve higher environmental standards cost-effectively'.[390] To this end, the UK government preferred to support industries and this became an important element of a major science policy initiative created during the 1990s: Technology Foresight (TF).[391]

In the early 1990s, representatives of the UK visited some other developed countries, including the Netherlands.[392] These countries employed so-called foresight policy activities, which were mainly based on widespread consultations among interested parties and individuals. Based on these activities, emerging science-based areas were then identified in order to receive governmental (and industrial) support. This support was expected to improve industrial competitiveness. But, these foresight activities required a broadly conceived and consensual style of policy making that went beyond isolated governmental committees. Furthermore, it could set in train the planning of an economy instead of harnessing market forces. Foresight activities therefore contrasted with UK's traditional methods of policy making.[393] This was clearly recognized by Sir William Stewart (Chief Scientific Adviser, UK Cabinet Office and Head of the Office of Science and Technology):[394]

> The UK's major competitors undertake Foresight, and we cannot afford to give them a competitive advantage by neglecting to take up the Foresight challenge.

In 1993, the UK government responded with the announcement of TF.[395]

According to the British government, this led in comparison with other countries to a type of foresight that was more a 'market-driven approach', in which future markets were first

[390] Ibid. 218.
[391] Office of Science and Technology, *Progress through Partnership: Report from the Steering Group of the Technology Foresight Programme 1995* (London: HMSO Publications, 1995), III-V.
[392] For example: Advisory Council on Science and Technology, *Technology Foresight: The Identification and Promotion of Emerging and Generic Technologies* (London: HMSO Publications, 1994).
[393] Ibid. See also: Pugh, *State & Society*; Jasanoff, *Designs on Nature*, 247-271; Luke Georghiou, "The UK Technology Foresight Programme," *Futures* 28 (1996): 359-377. Georghiou was a professor of science and technology policy and he was a director of support projects in TF.
[394] Office of Science and Technology, *Progress through Partnership*, V.
[395] Ibid., III; Anonymous, "UK Group to Explore Potential Benefits of 'Foresight' Exercise," *Nature* 373 (1995): 271.

identified and 'then the technologies, and related scientific research, which underpin them'.[396] This applied research approach was supported by pressures on the so-called 'conditions' of science, i.e., a downward trend of the public expenditures on R&D as a proportion of Gross Domestic Product (% GDP).[397] In practice, the first round of TF reflected this approach as far as the input of industry was concerned in the setting of research agendas. In the consultation processes related to the TF initiative of experts the input of industry dominated over academia and other interest groups.[398] Participants of TF stated that 'cultural constraints and constraints arising from general public attitudes or pressure groups' could form barriers to the production and acceptation of new knowledge.[399] The outcome of this broad policy initiative was that in the end six research themes were defined, including one related to chemistry, named 'A Cleaner World'.[400]

Some organizations embraced TF whereas others did not, or just partly.[401] Chemical sciences were certainly influenced by the 'rationale' of the TF initiative.[402] In the next section I will discuss how TF influenced the RSC: the research theme 'A Cleaner World' had a major impact on the rise of green chemistry within that organization.

3.4 Communicate (Green) Chemistry: a Scientific RSC Journal and a Network

During the 1990s, the RSC was confronted by a new science policy and the introduction of TF by the UK government. At the same time, the RSC experienced a negative public image of

[396] Office of Science and Technology, *Progress through Partnership*, 22; Georghiou, "The UK Technology Foresight," 361.
[397] Laurens K. Hessels and Harro van Lente, "Practical Applications as a Source of Credibility: a Comparison of Three Fields of Dutch Academic Chemistry," *Minerva* 49 (2011): 218. See also: Paul Stoneman, "Government Spending on Research and Development in the UK," *Fiscal Studies* 20 (1999): 223-259; UK Cabinet Office, *Realising our Potential: a Strategy for Science, Engineering and Technology* (London: HMSO Publications, 1993); Nick von Tunzelmann, "Technology in Post-war Britain," in *The Cambridge Economic History of Modern Britain*, Floud and Johnson, 305-307. See also: Thatcher, "Speech to the Royal Society".
[398] Georghiou, "The UK Technology Foresight," 367. The TF initiative reached out to over 10,000 people. See also: Barend van der Meulen, "The Impact of Foresight on Environmental Science and Technology Policy in the Netherlands," *Futures* 31 (1999): 9. The dominance of the industry was in line with the advices of the Advisory Council On Science and Technology. See heretofore: Advisory Council on Science and Technology, *Technology Foresight: The Identification*, 6.
[399] Georghiou, "The UK Technology Foresight," 375.
[400] Office of Science and Technology, *Progress through Partnership*, 80-81.
[401] Georghiou, "The UK Technology Foresight," 377. Georghiou argued that many organizations used TF as a benchmark for their own foresight.
[402] Hessels and Van Lente, "Practical Applications as a Source of Credibility," 218.

chemistry that was partly caused by news on chemical pollution.[403] Obviously, there were social and cultural obstacles for the acceptation of chemistry and chemical knowledge in the public domain. This was recognized by participants in the TF program for science at large. In this section I will show how the ideas and practices of TF started to influence the operations of the RSC. I will argue that important drivers for the recognition and emergence of green chemistry within the RSC were related to the policy context of TF, which also provided a matrix of new networking opportunities.

For environmental chemistry, the early 1990s was marked by the establishment of ECG.[404] This subfield of the chemical sciences continued with the key activities it had started during the 1970s and 1980s.[405] The ECG cooperated, for instance, with organizations such as the Association of Exploration Geochemists Environment Committee, with the Society for Environmental Geochemistry and Health, and with the interdivisional Toxicology Group, founded in 1987.[406] At the same time, some changes were taking place. According to Brendan Keely, the University of York started to offer courses that next to geochemistry and environmental analytical chemistry also included environmental modules that were based on catalysis. In Keely's view this was going to contribute to what he called "clean technology":[407]

> Environmental research in York covers a wide range of subject areas and many of the activities involve collaborative projects with industrial partners. The department has major interest in Clean Technology (led by Professor James Clark and supported by Duncan

---

[403] See Chapter 1.
[404] *Environmental Chemistry Group Newsletter* became the bulletin of ECG. Anonymous, "The Environmental Chemistry Subject Group," 1.
[405] See for example: Eric Davies, "The Environmental Chemistry Degree Course at Lancaster University," *Environmental Chemistry Group Newsletter* 2 (1995): 3-4; Roy Lowry, "Environmental Chemistry at Plymouth University," *Environmental Chemistry Group Newsletter* 3 (1996): 5-6; John Farmer, "Environmental Chemistry at the University of Edinburgh," *Environmental Chemistry Group Newsletter* 4 (1996): 5-7.
[406] Keith Nicholson, "News of UK – and USA – based Environmental Chemistry Groups," *Environmental Chemistry Group Newsletter* 2 (1995): 5; Holder, "Chairman's Report," 2. Membership numbers of the Toxicology Group leveled around 300, see heretofore annual reports and annual records of the RSC from 1988 onwards. Most likely, specialization regarding environmentally engaged chemical research increased, hereby leading to the institutionalization of different subject groups and membership competition among these groups. That's why in the late 1990s membership numbers of ECG slightly declined. See heretofore: "Our Divisions & Interest Groups," RSC, accessed May 22, 2022, http://www.rsc.org/membership-and-community/connect-with-others/through-interests/.
[407] Brendan Keely, "Environmental Chemistry at the University of York," *Environmental Chemistry Group Newsletter* 5 (1997): 3-4.

Macquarrie and Chris Rhodes) and is a leading exponent in the development of environmentally friendly catalysts to minimise waste in industrial processes and to avoid the use of toxic or undesirable solvents.

Clean technology was not at the fore within the environmental chemistry community in the mid-1990s, so the approach of the scientists at York was rather new.

The Scientific Forward Look initiated by the RSC did initially not include this clean technology approach (within ECG). In response to the declining R&D expenditures of the UK government, the Scientific Forward Look was an initiative of the RSC to counterbalance 'the "market pull" approach of Technology Foresight'.[408] RSC subject groups had the opportunity to influence RSC's science policy in order to determine the most important research topics for the future. The ECG presented its "ECG's Draft Scientific Forward Look Document" and recognized the monitoring and analysis of toxics, and their pathways, in the environment as fields of study for environmental chemists.[409] Also innovation in the analysis of particulate matter and volatile organic compounds, cheap field methods for 'low level concentrations' of toxics in aqueous environments, and cheap consistent 'toxicological assays' were mentioned in the document.[410] These future research topics were consistent with the dominant approaches within the ECG, as sketched in the previous section (3.2).[411] Clean technology was not part of these ideas within the ECG, nor was green chemistry on the agenda of the Group, contrary to the development of DEC in the USA in the same period.[412] Green chemistry entered the RSC via a different route.

In February 1995, a group of approximately fifty elite chemists from universities, industry and the RSC met to discuss TF at a workshop titled 'Chemistry in the UK: Will it Survive?'.[413] Under the leadership of Alan Calder (Chairman TF Chemicals Panel) and John

---

[408] RSC, *The Royal Society of Chemistry: Annual Review 1995* (London: RSC, 1996), 13.
[409] Anonymous, "The ECG's Draft Scientific Forward Look Document," *Environmental Chemistry Group Newsletter* 6 (1997): 2-3; RSC Occupational and Environmental Toxicology Subject Group, "The Royal Society of Chemistry's Scientific Forward Look: The Response of the Occupational and Environmental Toxicology Group," *Environmental Chemistry Group Newsletter* 7 (1998): 10-11.
[410] Anonymous, "The ECG's Draft Scientific Forward," 3.
[411] See also for example: Susan J. Parry, "Environmental Chemistry at The Centre for Analytical Research in the Environment, Imperial College," *Environmental Chemistry Group Newsletter* 6 (1997): 3-4; John Holder, "Environmental Chemistry at the University of Central Lancashire," *Environmental Chemistry Group Newsletter* 6 (1997): 6-7.
[412] See Chapter 2.
[413] RSC, *Chemistry in the UK: Will it Survive?; Conclusions of the Royal Society of Chemistry Workshops for Those in Universities and Industry Held on 13/14 & 20/21 February 1995* (London: RSC, 1995). This workshop was sponsored by the Department of Trade and Industry, Office of Science and Technology, Engineering and Physical Sciences Research Council (EPSRC) and the RSC.

Brophy (Vice-Chairman TF Chemicals Panel), several problems, prospects and chemistry-based research areas were identified that were relevant to TF. Among these problems was the negative public image of chemistry and the fact that chemistry in the eyes of the public 'lacks "green" credibility'.[414] This was considered a major obstacle in the further development of the field and removing that obstacle became a key recommendation of the TF Chemicals Panels to the steering group of TF.[415] Other recommendations encouraged partnerships that were centered on a wide range of chemical topics, including applied catalysis. These recommendations had a short-term interest. Long-term issues were formulated too, such as 'meeting environmental requirements'.[416]

While some of the elite chemists proposed actions on promoting new networks between academia and industry in order to support the market-oriented research of TF, others feared a diminishing of basic research.[417] This situation characterizes the ambivalent views concerning TF among leading UK chemists that influenced subsequent policies of RSC. This learned body was happy to be involved in TF, because as an interested stakeholder it could influence UK science & technology policy to a degree that was never realized before, according to RSC President Edward Abel.[418] The RSC recognized the importance of creating science-based networks in order to handle new knowledge demands, but which also accounted for the public that 'saw chemistry merely as a polluting and dangerous activity'.[419] With respect to TF the RSC reported in 1996:[420]

> Preliminary meetings were also held with the Natural Environment Research Council (NERC) aimed at identifying the relevance of chemistry to the NERC's remit, particularly in the key environmental areas recognized by the Technology Foresight programme.

The RSC thus realized that chemistry could contribute in solving environmental problems, and that new networks should be established. In the mid-1990s the RSC therefore organized so-called "technology car boot sales". These networking events were organized in the spirit of TF aiming for 'new businesses or research collaborations' such as 'an environmental

---

414 Ibid.
415 Office of Science and Technology, *Progress through Partnership*, 49.
416 Ibid., 49.
417 RSC, *Chemistry in the UK: Will it Survive?*.
418 Edward W. Abel, "A Message from the President," in RSC, *The Royal Society of Chemistry: Annual Review 1995*, 1.
419 Ibid., 15.
420 RSC, *The Royal Society of Chemistry: Annual Review 1996* (London: RSC, 1997), 13.

technology network'.[421] This took place in collaboration with major funding agencies, including EPSRC, NERC and the Biotechnology and Biological Sciences Research Council.[422] Within this new policy context – in which an environmental network was highly welcome and an improvement of the image of chemistry was on the agenda – the RSC Green Chemistry Network (GCN) was launched in 1998.[423]

GCN operated from York University under leadership of James Clark. In 1979, he had been appointed at York University, and was actively engaged in environmental research and education. Clark published dozens of journal articles and was appointed in 1994 as Chair in Industrial and Applied Chemistry by his university. Given his industrial and environmental interests he was able to obtain funding in the early 1990s with a Royal Academy of Engineering/EPSRC Clean Technology Fellowship.[424] That fellowship was part of the UK Clean Technology Programme, which started in 1992 and ran until 1997.

Clark used the term "clean technology" in his scientific communications. In 1996, he emphasized the importance of clean technology as a result of public scrutiny, economic considerations, and environmental quality standards that became increasingly tightened. In the same year, he defined clean technology in a review article:[425]

> Some of the major goals of clean technology in the chemical industry are to increase process selectivity, to maximise use of starting materials (aiming for 100 % atom efficiency), to replace stoichiometric reagents with catalysts and to facilitate easy separation of the final reaction mixture including the efficient recovery (and hopefully reuse) of the catalyst.

Clark was thus directly engaged with environmental issues related to the chemical industry.[426] Indeed, he participated in the 2nd Environmentally Benign Organic Synthesis meeting that was

---

421 RSC, *The Royal Society of Chemistry: Annual Review 1997* (London: RSC, 1998), 14.
422 Ibid.
423 This network was launched in advance of the second round of TF. For an overview of the second round, see: Steve Hanney, Mary Henkel and Dagmar von Walden Laing, "Making and Implementing Foresight Policy to Engage the Academic Community: Health and Life Scientists' Involvement in, and Response to, Development of the UK's Technology Foresight Programme," *Research Policy* 30 (2001): 12013-1219.
424 James H. Clark and Duncan J. Macquarrie, "Environmentally Friendly Catalytic Methods," *Chemical Society Reviews* 25 (1996): 303-310. The Clean Technology Programme was a small one compared to TF, but its chemical and contextual ideas were comparable to the ones of A Cleaner World. See also: Tim Lester, "Clean Technology Programme," *Green Chemistry* 1 (1999): G69-G71.
425 Clark and Macquarrie, "Environmentally Friendly," 303.
426 In fact, Clark argued that his environmental engagement principally started in the late 1980s. James H. Clark, (phone) interview, December 12, 2010.

held in August 17-22, 1997, at Queen's College (Oxford, the UK).[427] Clark lectured there about clean syntheses, and at the same conference Paul Anastas held his lecture 'Principles of Green Chemistry: Theory and Practice'.[428] The term green chemistry gradually became more well-known also in the UK. Clark was inspired by it and gave his own definition in 1998:[429]

> Some of the major goals of 'Green Chemistry' are to increase process selectivity, to maximise the use of starting materials (aiming for 100 % atom efficiency), to replace stoichiometric reagents with catalysts and to facilitate easy separation of the final reaction mixture including the efficient recovery of the catalyst.

Despite the change in vocabulary, he essentially gave the same definition as he did for clean technology in 1996.[430] So, interestingly, despite the rather different political contexts in the USA and the UK by 1998 chemists in both countries were starting to embrace the same environmentally related term: "green chemistry".

Next to the meeting at Queen's College, there were other influences from the USA that impacted the development of green chemistry in the UK. Brophy, with his engagement in networking as part of TF, argued that initiatives like the US Presidential Green Chemistry Challenge Awards and the GCI could contribute to a more environmentally friendly chemistry and would improve the public image of chemistry.[431] He also argued that pollution prevention was an accepted goal by UK governments, industries and society and chemists should contribute therefore to creating a sustainable society. This role for the chemical sciences was not recognized by the public and the UK government, as 'chemicals, chemistry and chemists are actually seen by many as a cause of the problems'.[432] He therefore proposed, with the support of Clark, that the RSC should 'take the initiative in the UK' to establish a 'Green

---

[427] This symposium was chaired by Stephen C. DeVito (EPA). Since 1998, this symposium has been titled "Green Chemistry". See: "Green Chemistry: Gordon Research Conference," Gordon Research Conferences: Frontiers of Science, accessed May 22, 2022, https://www.grc.org/green-chemistry-conference/default.aspx. In 1999, Clark chaired this symposium with an important green chemistry actor from the Netherlands: Roger Sheldon, who introduced the E factor. See Chapter 5.
[428] See for Paul Anastas Chapter 2 and Chapter 5.
[429] James H. Clark and Duncan J. Macquarrie, "Catalysis of Liquid Phase Organic Reactions Using Chemically Modified Mesoporous Inorganic Solids," *Chemical Communications* 8 (1998): 853.
[430] Clark, interview, December 12, 2010. Clark confirmed that, throughout the 1990s, he explored the same kind of activities in his laboratory, independently from the terms chosen.
[431] John Brophy, "A Major New Initiative in Environmental Chemistry: Proposal for a Royal Society of Chemistry Green Chemistry Network," *Environmental Chemistry Group Newsletter* 8 (1998): 2. Brophy wrote his text in May 1998.
[432] Ibid.

Chemistry Network' (GCN).[433] Brophy continued that GCN could 'promote awareness, education, training and practice of sustainable chemistry in the UK' and recommended that Clark should become its first director.[434]

That wake-up call was recognized soon by the RSC, as part of an activity at the interface between science and politics:[435]

> The Society [RSC] had it largest ever attendance at the Parliamentary Links Day on 23 June [1998] where presentations were made at the theme of "Environmentally Beneficial Chemistry". The topics covered included environment-friendly developments in coatings, fuel cells, paints, agrochemicals and acetic acid industrial processes, culminating in the launch of the Society's Green Chemistry Network.

In 1998 GCN was thus founded within RSC, but was located 'alongside the new Centre for Clean Technology' and 'funded by York University and the Wolfson Foundation, the University of York Science Education Group and the Chemical Industries Education Centre' under leadership of Clark.[436] RSC contributed by giving GCN a start-up grant for five years.[437] The RSC subsequently reported in its annual review:[438]

> The Green Chemistry Network (GCN), a major Society initiative, was established during the year. Its aim is to promote awareness and facilitate education, training and the practice of environmentally-beneficial chemistry in industry, academia and schools.

This resulted in concrete actions. Clark reported in *Chemistry in Britain* that GCN will 'publicise the benefits of sustainable development for chemists and the general public'.[439] Bob Hazell, in his turn, reported as the Secretary of the RSC Environment, Health and Safety Committee (EHSC) that the RSC informed the Department of Environment, Transport and the

---

[433] Ibid.
[434] See also: Peter O'Neill, "Initiatives on the Environment from the Royal Society of Chemistry," *Environmental Chemistry Group Newsletter* 8 (1998): 1; Clark, interview, December 12, 2010.
[435] RSC, *The Royal Society of Chemistry: Annual Review 1998* (London: RSC, 1999), 16. See also: James H. Clark, "From Brown Earth to Green Chemistry," *Science in Parliament* 55 (1998): 10. For clearness sake, the Parliamentary Links Day existed in advance of the John Major Administration and its TF.
[436] Brophy, "A Major New Initiative," 2-3. The EHSC also supported the launch of GCN.
[437] Clark, interview, December 12, 2010.
[438] RSC, *The Royal Society of Chemistry: Annual Review 1998*, 15. See also: John Brophy, "The Royal Society of Chemistry's Green Chemistry Network," *Environmental Chemistry Group Newsletter* 9 (1999): 16; James H. Clark, "Editorial," *Green Chemistry* 1 (1999): G1-G2.
[439] James H. Clark, "The Greening of Chemistry," *Chemistry in Britain* 34 (1998): 43-45.

Regions (DETR) about GCN, 'which aims to promote education and the practice of sustainable chemistry'.[440] Consequently, GCN prepared the UK's first green chemistry awards, which were announced in January 2000 in order to recognize the work of young academics and UK chemical companies.[441] These examples make clear that the term green chemistry had acquired a foothold within the RSC. The term was rhetorically useful in public dialogues, and was also instrumental in creating a political and research agenda.[442] From 1999 onwards the term green chemistry was increasingly used within scientific communications published by the RSC.

GCN was not a technical division like the ECG, but was closely connected to the chemistry department of York University. Clark, a dominant green chemistry advocate, was primarily an active laboratory chemist who co-produced, in line with the "operational realism" character of the discipline, new knowledge and substances.[443] Next to his GCN involvement as a scientist, he successfully campaigned for launching a new international journal, called *Green Chemistry* and that has been published by the RSC since its start in 1999.[444] This journal significantly contributed to the international output of articles that had "green chemistry" in their titles, keywords or abstracts (cf., Figure 3.2). During the years 2001-2005 the number of international journals that published articles using that term grew enormously, but *Green Chemistry* was able to keep its high share of about 20 percent of the total output.

---

440 Bob Hazell, "Sustainable Development: Submissions by the RSC's Environment, Health and Safety Committee to DETR Consultations on Aspects of Sustainable Development," *Environmental Chemistry Group Newsletter* 9 (1999): 14. Hazell was Secretary of the EHSC.

441 GCN cooperated with DETR, Salters Company, the Jerwood Charitable Foundation and the Department of Trade and Industry: RSC, *The Royal Society of Chemistry: Annual Review 1999* (London: RSC, 2000), 7. See also: Mike Lancaster, "UK Green Chemistry Awards," *Environmental Chemistry Group Newsletter* 11 (2000): 27; "Prizes Archive," RSC, accessed May 22, 2022, http://www.rsc.org/ScienceAndTechnology/Awards/GreenChemistry/index.asp.

442 GCN was also involved in the dissemination of a green chemistry publication for secondary schools in the UK. See: RSC, *The Royal Society of Chemistry: Annual Review 2001* (London: RSC, 2002), 7. In 2000, RSC was awarded the Green Apple Award. See: RSC, *The Royal Society of Chemistry: Annual Review 2000* (London: RSC, 2001), 7.

443 Bensaude-Vincent and Simon, *Chemistry: The Impure Science*, 6.

444 Brophy, "A Major New Initiative," 2-3; RSC, *The Royal Society of Chemistry: Annual Review 1999*, 14.

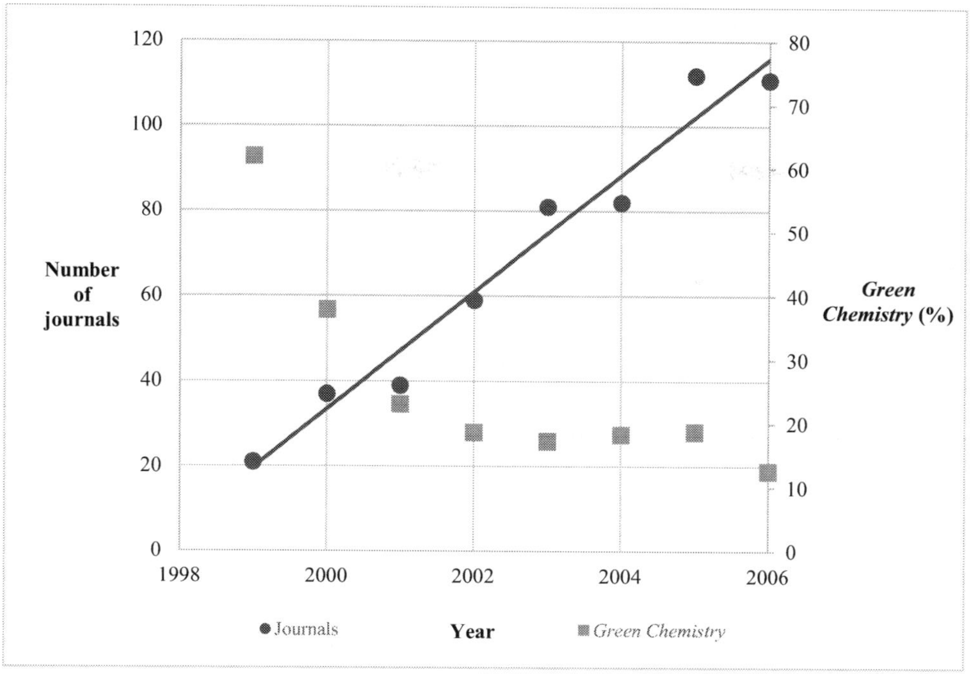

Figure 3.2: The linear growth in number of journals with green chemistry labeled articles and the percentage contribution in this respect of *Green Chemistry*.[445]

Clark became the first scientific editor of this journal. In his first editorial on the objectives and launch of *Green Chemistry* he wrote:[446]

> On the eve of the 21st century, however, the public are more aware of the hazardous substances that many chemical processes use and generate than the benefits of the products. Chemistry and the chemical industry have tarnished images. With your help we can use this journal to convey the means and motivation for chemists to make a difference.

This viewpoint was in line with one of the major motives why the RSC supported the GCN: its concerns about the negative public image of chemistry.[447]

---

445 Data obtained from ISI Web of Knowledge: Web of Science (accessed August 4, 2020), by searching the topic "green chemistry".
446 Clark, "Editorial," G2.
447 See also: Edward W. Abel, "Good News from Chemistry: Meeting the Environmental Challenge," *Science in Parliament* 55 (1998): 2-3. Most likely, RSC expected good revenues from *Green Chemistry*.

Although Clark used with "clean technology" another environmentally related term during the mid-1990s, he preferred the term green chemistry since 1998. With respect to the title of his new journal he commented that:[448]

> The title *Green Chemistry* was itself the subject of considerable discussion and debate. We considered many alternatives, but none carried the same combination of widespread use and appreciation, as well as simplicity and impact.

On the one hand, it illustrated that the term green chemistry was increasingly used by 1999. On the other hand, it shows that there were still opposing views concerning the term. In 1999, Mike Lancaster (Network Manager of GCN) specified the influences from the USA:[449]

> The term Green Chemistry is rather emotive, intentionally so, and conjures up different meanings to different people. Its origins are with the US Environmental Protection Agency, but the movement is rapidly spreading across Europe and is creating a great deal of interest elsewhere.

Clark's editorial and Lancaster's reflection upon the term green chemistry illustrate that there was some ambivalence in the recognition of green chemistry. On the one hand, the identity and the internal scientific meaning of green chemistry were not totally clear. This will further be discussed in Chapter 5.[450] On the other hand, although Clark experienced a lack of recognition to obtain submissions in the first years of *Green Chemistry*, his journal developed in a very positive way.[451] The readership of *Green Chemistry* grew considerably.[452] Therefore, RSC was satisfied with the results obtained by *Green Chemistry*. From 1999 up to and including 2003 this journal released six issues per year. This changed in 2004 with the appointment of Walter Leitner as the new scientific editor, because then the publication

---

[448] Clark, "Editorial," G1.
[449] Mike Lancaster, "Further News of the Royal Society of Chemistry's Green Chemistry Network," *Environmental Chemistry Group Newsletter* 10 (1999): 13.
[450] Clark, interview, December 12, 2010. Clark confirmed that some people had problems with the term green chemistry.
[451] Ibid.; James H. Clark et al., "15 Years of *Green Chemistry*," *Green Chemistry* 16 (2014): 18-23.
[452] The readership of Clark his own works also grew. (Source: ISI Web of Knowledge: Web of Science.) This is in line with a theoretical account on the position of editorship, see: Lydia L. Lange and P.A. Frensch, "Gaining Scientific Recognition by Position: Does Editorship Increase Citation Rates?," *Scientometrics* 44 (1999): 459-486.

frequency increased to twelve issues per year.[453] Currently, this frequency is 24 issues per year.[454]

3.5 Discussion: UK's Road to Green Chemistry

Public environmental concerns gained political momentum around 1970, as witnessed by new environmental legislations and the establishment of RCEP in 1970. This momentum weakened during the Thatcher Administration, especially during years of economic crisis. Starting in the late 1980s, public environmental concerns intensified again, though. The Thatcher Administration was at first reluctant to regulations and environmental policies that interfered with market forces, but later new environmental policies were developed as a response to EEC pressures and also probably for electoral reasons. These new policies included traditional features of UK's communitarian civic epistemology. Despite the tightening of emission standards for hazardous compounds and an increase in prosecutions, the legal and institutional reforms from the late 1980s onwards regarding environmental policy were more evolutionary than revolutionary.[455] The close cooperation with industry in the political tradition of the UK remained, though with an increase in governmental transparency and open dialogue.[456]

The growing public environmental concerns and the political attention paid to it went hand in hand with the national demand for knowledge concerning the impact and pathways of chemicals in the environment. This is shown by the rising membership of environmental chemists within the RSC (Figure 3.1) from the late 1980s onwards. Environmental chemistry was fully institutionalized within the RSC in 1994.

---

453 Walter Leitner, "Green Chemistry: a New Phase," *Green Chemistry* 6 (2004): G1. The number of pages also increased from 292, in 1999, to 616, in 2004.
454 "Green Chemistry," RSC, accessed May 22, 2022, http://www.rsc.org/journals-books-databases/about-journals/green-chemistry/.
455 Connelly and Smith, *Politics and the Environment*, 304-307; Georgina Crowhurst and Jasmine Murphy, "UK's Environmental Watchdog Brought to Heel," *European Environmental Law Review* 15 (2006): 234-237. For example, the HMIP West Division reported 4 prosecutions in its 1991-92 review, with fines typically in an order of magnitude of £ 1,000, and 12 people from the public who consulted the HMIP's registers. See: HMIP, *Her Majesty's Inspectorate of Pollution Annual Report 1991-92: West Division Review* (London: HMSO Publications, 1993), 8 and 23-24.
456 Connelly and Smith, *Politics and the Environment*, 304-307; Gemmill, "The Approach of Her Majesty's Inspectorate of Pollution," 3-4; Emmott and Haigh, "Integrated Pollution Prevention and Control," 301-311. See also: HMIP, *Her Majesty's Inspectorate of Pollution Annual Report 1991-92*, 9.

Moreover, shaped by environmental concerns, the UK profession of chemistry experienced a negative public image. Applying the 'science-society contract' idea proposed by Laurens Hessels and Harro van Lente, this affected the (presumed) 'identity' of chemistry and thereby the possibilities to finance chemical research.[457] With the nation's industrial competitiveness in mind, the UK government responded with new science and technology policies, most notably the launch of TF in 1993. The policy style of TF was new for the UK, as interested parties were more and broadly involved than ever and weakened hereby the power of isolated governmental committees. On the other hand, with the decline in government R&D expenditures and the described market oriented approach, TF also mirrored traditional elements of UK's political culture.

As a learned and professional body, RSC embraced TF in an ambivalent manner thereby searching for new policy directions as witnessed by its Scientific Forward Look. This in turn led to a changing culture of opportunities within the RSC, which demanded for a new network in order to address new knowledge demands, such as pollution prevention. In that new context, the strategic launch of GCN and *Green Chemistry* took place during the late 1990s within the RSC. In part, these actions were responses to green chemistry initiatives in the USA.

GCN and *Green Chemistry* were coordinated by academic research chemists such as Clark from York University. He and his colleagues were close to the laboratory, and without daily institutional or policy influences from the UK government. The negative image of chemistry provided a strong motive also for Clark and the leaders of the RSC. That favored the growth and recognition of green chemistry in the UK.[458] Green chemistry was considered a convenient rhetorical tool to improve the public image of chemistry. Because as a metaphor, it was used in the communication with society and politicians. At the Parliamentary Links Day, for instance, it was made clear that chemists were engaged with the problems of the environment.[459] Within the circulation of chemical knowledge, the term green chemistry

---

[457] Hessels and Van Lente, "Practical Applications as a Source of Credibility," 215-240. See also Chapter 1.

[458] Diana Crane, "Fashion in Science: Does it Exist?," *Social Problems* 16 (1969): 433-441.

[459] For a theoretical framework on metaphors in science communication, see: Vasilia Christidou, Kostas Dimopoulos and Vasilis Koulaidis, "Constructing Social Representations of Science and Technology: the Role of Metaphors in the Press and the Popular Scientific Magazines," *Public Understanding of Science* 13 (2004): 347-362. See also: Jesús P.Z. Bonilla, "Rhetoric, Induction, and the Free Speech Dilemma," *Philosophy of Science* 73 (2006): 175-193. Following Bonilla, rhetoric is the strategic use of language.

increasingly became a commonly-used notion in the international language of science communication too (cf., Figure 3.2).

As a leading man in the development of green chemistry in the UK, Clark experienced a reverse to the medal of the growth of green chemistry, namely the lack of good quality submissions to his scientific journal, in particular during the first years.[460] This hints at another type of recognition; a scientific one among academic top-chemists. This vulnerability regarding the field of green chemistry will be discussed in Chapter 5.

---

[460] Clark, interview, December 12, 2010.

Chapter 4: The Rise of Green and Sustainable Chemistry in the Netherlands

After my analysis of the emergence, recognition and institutionalization of green chemistry within the national chemical societies of the USA and the UK, I will now focus on the rise of green and sustainable chemistry in the Netherlands, and the role of the Royal Netherlands Chemical Society [Koninklijke Nederlandse Chemische Vereniging] (KNCV) therein.

In the first section I will discuss the political and cultural attention paid to environmental problems in the Netherlands during the 1960s-1980s. Thereafter, the emphasis shifts to the KNCV, a professional and learned body with ca. 14,000-16,000 members during the 1990s, who worked in industry, universities, government, and education.[461] The impact of environmental concerns for the KNCV is discussed in that section together with the rise of environmental chemistry as a new subfield of chemical science in the Netherlands. In a third section, similar to the cases of the USA and the UK, I will analyze the rise of pollution prevention policies and the formation of new networks among various actors within the Dutch chemical arena. I will argue that it is mainly from these new networks that fields of study labelled as green or sustainable chemistry emerged. In order to understand these developments, special attention will be paid to changing perceptions and values with respect to the state of the environment, and to changes in the public image of chemistry. The key research questions of this chapter therefore are: what and who are the dominant factors and actors that influenced the emergence and recognition of green and sustainable chemistry, and the associated vocabularies, in the Netherlands? And why were these factors and actors dominant?

4.1 Environmental Concerns and Government Action in Dutch Society: 1960s-1980s

Similar to elsewhere in the Western world the attention paid to environmental issues changed greatly in the Netherlands during the decades following World War II.[462] In the course of the

---

[461] Currently, the KNCV has about 8,500 members. See: "Over de KNCV," KNCV, accessed May 22, 2022, http://www.kncv.nl/vereniging/overzicht. The KNCV was founded in 1903 as the 'Algemene Nederlandsche Chemische Vereeniging'.

[462] See for example: Jan L. van Zanden and Simon W. Verstegen, ed., *Groene Geschiedenis van Nederland* (Utrecht: Het Spectrum, 1993); Henk van Zon, *Tachtig Jaar RIVM* (Assen: Van Gorcum, 1990); Arthur P.J. Mol, *The Refinement of Production: Ecological Modernization Theory and the*

1950s there was a strong focus on stimulating industrialization and reducing unemployment, even if this took place at the expense of environmental quality. Together with continued economic growth, the 1960s witnessed a changing social-cultural climate, hand in hand with an increase in democratization and a lifting of sociopolitical barriers.[463] Parallel to this, also environmental pollution received growing public attention, partly because of the international impact of Rachel Carson's book on the ecological dangers of pesticides, and partly because of the growing impact of emissions caused by traffic, agriculture and industries on the quality of air and water.[464] New action groups and committees were therefore founded that focused on ecological damage and dangers to public health by air pollution in industrial areas.[465] For instance, the Association Against Air Pollution [Vereniging Tegen Luchtverontreiniging] was founded in 1963 in the highly industrialized Rotterdam area, and in 1972 the national Foundation for Nature and the Environment [Stichting Natuur & Milieu] was established.[466] The government responded to these public voices by creating agencies that should take care for the environment and for public health. An early example is the Inspectorate of Environmental Health [Inspectie Milieuhygiëne] (IM), created in 1962.[467]

In the 1970s, environmental consciousness peaked, together with new state regulations that partly ran parallel to international events such as the 1972 United Nations Conference on the Human Environment in Stockholm.[468] In 1970, the Surface Waters Pollution Act [Wet

---

Chemical Industry (Utrecht: Van Arkel, 1995), 119. For political perspectives, see: Jac Bosmans and Alexander van Kessel, *Parlementaire Geschiedenis van Nederland* (Amsterdam: Boom, 2011).

463 See for example: Jan J. Woltjer, *Recent Verleden: De Geschiedenis van Nederland in de Twintigste Eeuw* (Amsterdam: Uitgeverij Balans, 1992); Dirk P. Blok et al., ed., *Algemene Geschiedenis der Nederlanden; Deel 15* (Haarlem: Fibula-Van Dishoeck, 1982).

464 Kees Boender, *Milieuprotest in Nederland: Sociologische Analyse van Milieusolidariteit onder Elites en Publiek* (Rijswijk: Sijthoff Pers, 1985), 531-532; Jan L. van Zanden, "Omstreden Landbouw," in *Groene Geschiedenis*, Van Zanden and Verstegen, 76-78; Rachel Carson, *Silent Spring* (New York, NY: Houghton Mifflin Company, 1962). Carson also paid attention to the impact of pesticides on public health.

465 Van Zanden, "Omstreden Landbouw," 78; Jan L. van Zanden, "De Veranderende Relatie tussen Mens en Natuur," in *Groene Geschiedenis*, Van Zanden and Verstegen, 193-198.

466 See: Mol, *The Refinement of Production*, 126; Jacqueline M. Cramer, *De Groene Golf: Geschiedenis en Toekomst van de Milieubeweging* (Utrecht: Uitgeverij Jan van Arkel, 1989), 73-78; Boender, *Milieuprotest in Rijnmond*, 216-221, 268-276 and 531.

467 Van Zon, *Tachtig Jaar*, 256. See for the announcement of IM in the Dutch House of Representatives: Tweede Kamer der Staten-Generaal, "Kamerstuk 6100 XII, ondernummer 22: Verslag van het Mondeling Overleg Betreffende Afdeling VII," 1961. The Dutch House of Representatives is, in the Dutch language, known as the "Tweede Kamer".

468 Nico Nelissen, Jan van der Straaten and Leon Klinkers, ed., *Classics in Environmental Studies: An Overview of Classic Texts in Environmental Studies* (Utrecht: International Books, 1997); Juan C. di Primio, "Data Quality and Compliance Control in the European Air Pollution Regime," in *The Implementation and Effectiveness of International Environmental Commitments: Theory and Practice*, ed. David G. Victor et al. (Cambridge, MA: the MIT Press, 1998), 285.

Verontreiniging Oppervlaktewateren] became effective, and also the Air Pollution Act [Wet inzake de Luchtverontreiniging], which mostly concerned industrial emissions, such as that of SO_2.[469] In 1971, the Ministry of Public Health & Environmental Health [Ministerie van Volksgezondheid & Milieuhygiëne] was established in order to regulate emissions, and minimize the impact of pollution on public health.[470] Based on these new environmental laws and the creation of new Dutch government agencies the control of pollution became more centralized in the Netherlands, and the activities of governments regarding industrial emissions became more transparent for the public at large.[471] In response to the recommendations of the Stockholm conference, the Dutch government issued in 1972 the Urgency Report Environmental Health [Urgentienota Milieuhygiëne] aiming at more stringent environmental quality standards, and at an improvement of the scientific understanding of the environment.[472] Despite these government initiatives, control of environmental pollution to serve the public good was still considered weak by most of the environmental action groups.[473]

The public call for environmental control was strongest in densely populated areas where industrial pollution was significant, such as the Rijnmond area, near Rotterdam, which was governed by the Rijnmond Public Authority [Openbaar Lichaam Rijnmond].[474] As a result of growing pressure from society, that government body decided in 1971 to create a new department: the Central Environmental Protection Agency for the Rijnmond area [Dienst Centraal Milieubeheer Rijnmond] (DCMR).[475] Louis Clarenburg, one of the environmental

---

469 Gerda Dinkelman, *Verzuring en Broeikaseffect. De Wisselwerking tussen Problemen en Oplossingen in het Nederlandse Luchtverontreinigingsbeleid (1970-1994)* (Utrecht: Uitgeverij Jan van Arkel, 1995), 38 and 69.

470 For an overview of changing political consciousness with respect to environmental legislation in the Netherlands, see: Simon W. Verstegen, "Stedelijke en Industriële Vervuiling," in *Groene Geschiedenis*, Van Zanden and Verstegen, 146-149; Cramer, *De Groene Golf*; Bertien Broekhans, *Hoe Milieukunde Geschiedenis Werd: Demarcatie van een Maatschappelijk Relevante Wetenschap 1970-2000* (Ede: Nijmegen University Press, 2003).

471 Until the 1970s, local governments were principally responsible for the control of pollution caused by firms that were in conflict with individuals. Boender, *Milieuprotest in Rijnmond*, 269; Verstegen, "Stedelijke en Industriële Vervuiling," 146-149.

472 L.B.J. Stuyt, "Urgentienota Milieuhygiëne," letter of the Minister of Public Health & Environmental Health to the members of the Dutch House of Representatives, July 4, 1972; Ministerie van Volksgezondheid & Milieuhygiëne, "Kamerstuknummer 11906, ondernummer 2: Urgentienota Milieuhygiëne," July 4, 1972.

473 Van Zanden, "De Veranderende Relatie tussen Mens en Natuur," 193-198.

474 Boender, *Milieuprotest in Rijnmond*, 532. See also: Louis A. Clarenburg, "Penalization of the Environment Due to Stench: A Study of the Perception of Odorous Air Pollution by the Population," *Atmospheric Environment* 7 (1973): 333-351.

475 Boender, *Milieuprotest in Rijnmond*, 532. Today, DCMR is still an important controlling agency that also advises other public bodies.

authorities in the area, wrote that in 1970 the Rijnmond Authority received over 17,000 complaints of citizens concerning industrial stench, and in 1971 almost 30,000.[476] This significant increase in a short period of time illustrates both the growth of air pollution and the growing public concerns with respect to environmental health in 'one of the most impressive industrial agglomerations in the world'.[477]

As a result of the oil crisis of 1973 and the following economic depression of the mid-1970s, public concerns shifted from the environment to the economy, in particular to rising unemployment figures (cf., Figure 4.1).[478] Membership numbers of environmental NGOs stagnated, but this changed again in the course of the 1980s due to growing economic prosperity and to the rise of new environmental issues such as soil pollution and climate issues such as ozone layer depletion.[479] The environment shifted more and more from the periphery to the center of the political agenda during the 1980s, with also a government that strived for improved environmental control.[480]

---

476 Clarenburg, "Penalization of the Environment Due to Stench," 334.

477 Ibid. See: Ministerie van Volksgezondheid & Milieuhygiëne, "Urgentienota Milieuhygiëne," 8 and 36-38. See also: F. Langeweg, ed., *Zorgen voor Morgen: Nationale Milieuverkenning 1985-2010* (Alphen aan de Rijn: RIVM, 1988), 85. This report confirms the huge industrial production and energy consumption per km^2 of the Netherlands as compared to other industrialized economies, e.g., USA and the UK.

478 Van Zanden, "De Veranderende Relatie tussen Mens en Natuur," 196-199.

479 Ibid.

480 Verstegen, "Stedelijke en Industriële Vervuiling," 146-149. The view of Verstegen is in line with the theory of the environmental sociologists Dunlap and Mertig who emphasize a causal relationship between the economic tide and public environmental concerns. See: Riley E. Dunlap and Angela G. Mertig, "The Evolution of the U.S. Environmental Movement from 1970 to 1990: an Overview," in *American Environmentalism: the U.S. Environmental Movement, 1970-1990*, ed. Riley E. Dunlap and Angela G. Mertig (Washington, DC: Taylor & Francis New York, 1992), 1-10. See also: Van Zanden, "De Veranderende Relatie tussen Mens en Natuur," 194.

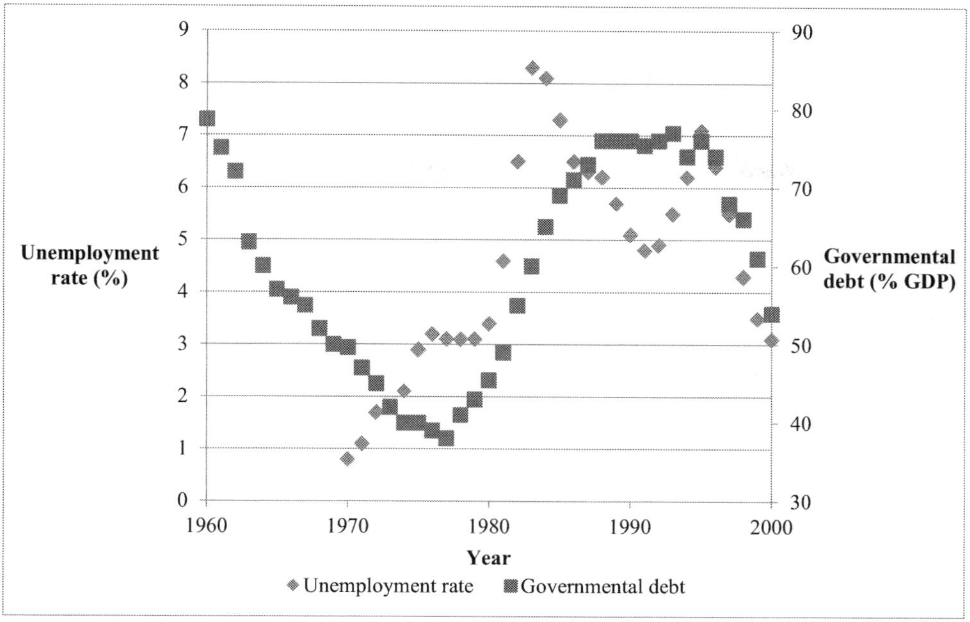

Figure 4.1: Economic key numbers of the Dutch economy for the 1960-2000 period.[481]

During the late 1970s and early 1980s environmental concerns regarding air and water, as well as local waste disposal, received increased public attention.[482] Also transboundary air pollution received much international public attention due to the negative impact of acid rain on European forests. The Dutch government signed the multilateral *1979 Convention on Long-Range Transboundary Air Pollution* (*LRTAP*).[483] Pollution was thus not solely a public health concern, but also became more and more an international ecological one.[484] This changing perception concerning pollution reached government circles in the early 1980s through the pressure of environmental NGOs that undertook and supported scientific

---

481 These data are obtained from the CPB Netherlands Bureau for Economic Policy Analysis at https://www.cpb.nl/raming-november-2020-vooruitzicht-2021#cijfers, accessed May 22, 2022. The exact unemployment data for the period 1960-1969 were not available, but they were low.

482 According to the environmental scientist Jacqueline Cramer, approximately 8000 places in the Netherlands were polluted. She reports a list of accidents concerning waste disposal sites: Cramer, *De Groene Golf*, 90-91.

483 For an overview of *LRTAP*, see: Pamela S. Chasek, David L. Downie and Janet W. Brown, *Global Environmental Politics* (Philadelphia, PA: Westview Press, 2010), 121-129; United Nations, *Convention on Long-Range Transboundary Air Pollution* (Geneva: United Nations, 1979). In United Nations Treaty Collection; available from https://treaties.un.org/Pages/ViewDetails.aspx?src=TREATY&mtdsg_no=XXVII-1&chapter=27&clang=_en, accessed May 22, 2022; Dinkelman, *Verzuring en Broeikaseffect*, 62-63.

484 Dinkelman, *Verzuring en Broeikaseffect*, 35-97.

investigations on the impact of acid rain on Dutch forests, as well as on the exterior of historic buildings.[485] In 1982 the Ministry for Housing and Physical Planning was renamed and reorganized into the Ministry for Housing, Physical Planning and the Environment (VROM) in order to develop more integrated policies concerning the state of the environment and physical planning in the densely populated and highly industrialized Netherlands.[486] Furthermore, the Directorate-General Environmental Protection was created within VROM under active leadership of the Minister of VROM, Pieter Winsemius.

As pointed out by historian Gerda Dinkelman, Winsemius initially received a lot of distrust from environmental NGOs and left-wing politicians who suspected him, being a member of the economic liberal party, to solely serve the industrial interests.[487] Winsemius realized this, and argued in a retro-perspective that solving environmental problems was preceded by a 'level of disagreement' among stakeholders.[488] He therefore also took the concerns expressed by the environmental NGOs on board, and in the early 1980s fully recognized acid deposition as a major policy challenge.[489] As a result, the government plans for the years 1985-1989 contained environmental quality standards that were more strict than those of other countries of the European Economic Community (EEC).[490]

Since the early 1980s the Netherlands was one of the 'lead states' of the *LRTAP* initiative.[491] The first Protocols of *LRTAP* concerned a 30 % reduction in SO_2 emissions (Helsinki Protocol, 1985), and the freezing in emission levels of NO_x (Sofia Protocol, 1988). Both were signed by the Netherlands, even though the country was a net exporter of SO_2.[492] The Netherlands also intended to bring their new environmental policies in line with those of

---

[485] Ibid., 84-90; Van Zanden, "De Veranderende Relatie tussen Mens en Natuur," 199-204.
[486] Bosmans and Van Kessel, *Parlementaire Geschiedenis*, 165-172; Dinkelman, *Verzuring en Broeikaseffect*, 89-90 and 100; Tweede Kamer der Staten-Generaal, "Kamerstuk 17555, ondernummer 7: Kabinetsformatie 1982," October 30, 1982, 5 and 76-77.
[487] Dinkelman, *Verzuring en Broeikaseffect*, 89-90 and 99-100.
[488] Pieter Winsemius, "Opening Address," *The Science of the Total Environment* 45 (1985): 7-15.
[489] Dinkelman, *Verzuring en Broeikaseffect*, 90 and 110-119.
[490] Ibid., 88, 94-96 and 110-119; Tweede Kamer der Staten-Generaal, "Kamerstuk 18605, ondernummer 2: Indicatief Meerjaren Programma Lucht 1985-1989," September 18, 1984, 5-10 and 51-53. See also: Graham Bennett, "The History of the Dutch National Environmental Policy Plan," *Environment* 33 (1991): 7-9 and 31-33.
[491] Chasek, Downie and Brown, *Global Environmental*, 121-129. *LRTAP* countries such as Sweden, Denmark and the Netherlands took the lead in advocating strict environmental quality standards in international negotiations. See Chapter 2 and 3 for the USA and the UK. The Netherlands increasingly used natural gas (low sulfur content) for economic reasons, which supported the reduction of Dutch SO_2 emissions. Frank Notten, *De Nederlandse Economie: De Invloed van de Aardgaswinning op de Nederlandse Economie* (Den Haag: Centraal Bureau voor de Statistiek, 2015), 3-5.
[492] Dinkelman, *Verzuring en Broeikaseffect*, 49-55 and 62-63.

neighboring countries, such as the German Federal Republic, that had introduced stringent environmental quality standards.[493]

Changes in environmental policy during the 1980s resulted by 1990 to a Dutch government expenditure on environmental R&D that was relatively high as compared to other European countries.[494] A new Cabinet also announced in 1990 that in case of environmental violations, government would act more firmly than before.[495] Administrative actions and prosecution by DCMR of (chemical) companies that violated environmental legislation also increased at the regional level.[496] All these government actions and new legislation were also a response to growing environmental concerns among the population. Membership figures and sponsoring of environmental organizations grew as well. In 1980 Greenpeace, for instance, had 18,000 members and sponsors in the Netherlands, but in 1990 this had grown to 600,000.[497] Although complaints about stench and smog also grew, paradoxically atmospheric concentrations of hazardous compounds strongly decreased during the 1980s.[498]

The Association of the Dutch Chemical Industry [Vereniging van de Nederlandse Chemische Industrie] (VNCI), which organized the collective interests of Dutch chemical industries, frequently complained during the 1980s that the Dutch environmental rules increasingly became more strict than those of other (European) countries. This would weaken the competitive position of the Dutch chemical industry.[499] Therefore, the VNCI advised the Dutch government that environmental quality standards should be harmonized with EEC

---

[493] Ibid., 115-119; Tweede Kamer der Staten-Generaal, "Kamerstuk 18605, ondernummer 2: Indicatief Meerjaren Programma Lucht 1985-1989," 59-60.

[494] See for example: J.M.M. Ritzen, "Wetenschapsbudget 1991: Kamerstuk 21805," letter of the Minister of Education & Science to the Dutch House of Representatives, September 18, 1990, 188.

[495] J.G.M. Alders (Minister of VROM) and J.R.H. Maij-Weggen (Minister of Transport, Public Works and Water Management), "Nationaal Milieubeleidsplan," letter to the members of the Dutch House of Representatives, May 28, 1990. See also in this respect: DCMR, *Jaarverslag 1990* (Schiedam: DCMR, 1991), 3. To be clear, "jaarverslag" is Dutch for "annual report".

[496] DCMR, *Jaarverslag 1990*, 62 and 63-86. This report lists companies that were subjected to strict environmental control in the form of administrative actions and criminal prosecution.

[497] SCP, *Sociaal en Cultureel Rapport 1998: 25 Jaar Sociale Verandering* (Rijswijk: SCP, 1998), 757-758. This report outlines an average increase in membership of no less than '510 %' for the largest Dutch environmental NGOs for the 1980-1997 period.

[498] DCMR, *Jaarverslag 1990*, 16 and 27-28. See also: United Nations Economic Commission for Europe, *Updated Handbook for the 1979 Convention on Long-range Transboundary Air Pollution and its Protocols* (Geneva: United Nations, 2015), 110. This report outlines Dutch SO_2 emissions for 1980 (466 kilotons per year) and 1990 (207 kilotons per year), which is a 55.6 % decrease.

[499] VNCI, *Jaarverslag 1980* (Den Haag: VNCI, 1981), 2; VNCI, *Jaarverslag 1984* (Den Haag: VNCI, 1985), 22; VNCI expressed its concerns about quality standards for several substances, e.g., chlorofluorocarbons (CFCs) and carbon mono-oxide. See: VNCI, *Jaarverslag 1985* (Den Haag: VNCI, 1986), 18-19.

guidelines.[500] Already during the early 1980s the VNCI proposed a dialogue with government, leading to mutual understanding and a 'fruitful cooperation'.[501] This led in 1986 to a pilot-project for the development of an environmental enforcement model in which Dutch government, and the VNCI participated.[502] It was presented by VNCI President Ruud Selman as a demonstration that the chemical industry during the 1980s had developed a 'constructive' dialogue with governments, instead of being trapped in a polarized interaction.[503] The dialogue that resulted was a presage of the pollution prevention policies of the 1990s.

4.2 The Emergence of Environmental Chemistry in the Netherlands

Many of the public concerns sketched in the previous section were related to chemical industries, and chemical hazards. Of course this went on not unnoticed by chemists and by their national chemical society, the KNCV. Many chemists were also motivated to contribute to a cleaner environment. In this section I will investigate how the growing public and governmental attention paid to the environment and public health influenced the work of chemists and the KNCV.[504] I will focus in particular on a new division that was created in 1970 within the KNCV, the Division Environmental Chemistry [Sectie Milieuchemie] (hereafter MC). I use the study of the MC as an instrument to map the emergence of environmental chemistry in the Netherlands, as well as the factors that impacted its further development.

---

500 VNCI, *Jaarverslag 1985*, 19. According to the VNCI, the EEC guidelines should be based on the guidelines proposed by the European Chemical Industry Council [Conseil Européen des Fédérations de l'Industrie Chimique]. See also: P.H. Bottelberghs, "Risk Analysis and Safety Policy Developments in the Netherlands," *Journal of Hazardous Materials* 71 (2000): 59-84. The international Organisation for Economic Co-operation and Development also pleaded for harmonization of guidelines. See: Taskforce KNCV, *Milieu: Chemici een Zorg* (Den Haag: KNCV, 1990), 8-15. This taskforce pleaded for harmonization with EEC guidelines, which should be scientifically based. Parallel to this, a growing attention was also paid by the VNCI to chemical waste problems. See: VNCI, *Jaarverslag 1981* (Den Haag: VNCI, 1982), 1.

501 VNCI, *Jaarverslag 1981*, 6 and 17. All translations of quotations in this chapter are mine, unless otherwise stated.

502 VNCI, *Jaarverslag 1986* (Den Haag: VNCI, 1987), 7. And in 1988, the VNCI was represented in project groups of VROM that concerned the prevention of halogenated organic waste. See: VNCI, *Jaarverslag 1988* (Den Haag: VNCI, 1989), 16.

503 R.E. Selman, preface to VNCI, *Jaarverslag 1989* (Den Haag: VNCI, 1990), 3.

504 Laurens K. Hessels and Harro van Lente, "Practical Applications as a Source of Credibility: a Comparison of Three Fields of Dutch Academic Chemistry," *Minerva* 49 (2011): 215-240.

The establishment of MC was prepared in 1969 by a steering group headed by the analytical chemist Geo Dijkstra.[505] Under the leadership of Dijkstra, articles of association were formulated and the first board was constituted in 1970 by members who worked in government, industry and academia. The MC was supposed to organize symposia, in collaboration with other interest groups, but also to serve Dutch society by contributing to a cleaner environment. After some debate, the members of MC preferred the term "environmental chemistry" as name of the Division.[506] This term emphasized the relationship between the environment and chemistry and was in line with the terminology employed in international academic circles.[507] The new term was defined at the start of MC as follows:[508]

> The research on chemical ingredients of the biosphere, both natural and synthetic ones; the research of chemical processes in the biosphere; and the investigation of chemical engineering processes that are in agreement with the standards required for a healthy environment; thereby aiming for the creation of optimal conditions for man, animal and plant.

With such a broad definition, environmental chemistry became, from a chemical point of view, a heterogeneous subfield, which included the determination of trace metals in biological systems by Dijkstra and his coworkers, but also many other topics.[509] Clarenburg, who was also one of the first board members of MC, for instance investigated air pollution caused by chemical industries. In 1971, he published a statistical analysis of a great amount of volumetric SO_2 samples.[510] Sundström and Hutzinger, by contrast, employed sophisticated

---

[505] Stuurgroep Sectie Milieuchemie, *Koninklijke Nederlandse Chemische Vereniging: Sectie Milieuchemie*, August 4, 1970; KNCV, "Jaarverslag van de Koninklijke Nederlandse Chemische Vereniging over 1970," *Chemisch Weekblad* July 23/30 (1971): 28. Dijkstra also chaired the KNCV in 1976. See: Ernst Homburg and Lodewijk Palm, ed., *De Geschiedenis van de Scheikunde in Nederland 3: De Ontwikkeling van de Chemie van 1945 tot het Begin van de Jaren Tachtig* (Delft: Delft University Press, 2004), 333.

[506] Stuurgroep Sectie Milieuchemie, *Koninklijke Nederlandse Chemische Vereniging*, 1.

[507] See also the Chapters 2 and 3.

[508] [De studie van chemische bestanddelen van de biosfeer, zowel van natuurlijke als kunstmatige origine; de studie van chemische processen in de biosfeer; de studie van de proceschemie, gericht op de uit de milieuverzorging voortkomende eisen; één en ander in het bijzonder gericht op het scheppen van optimale levenscondities voor mens, dier en plant.] As in: Stuurgroep Sectie Milieuchemie, *Koninklijke Nederlandse Chemische Vereniging*, 2.

[509] J.K. Terlouw, J.J. de Ridder, W. Heerma and G. Dijkstra, "Mass-spectrometric Determination of Metal Chelates," *Fresenius' Zeitschrift für Analytische Chemie* 249 (1970): 296. Dijkstra employed analytical methods, e.g., gas chromatography and infrared spectroscopy. See: Leo de Galan, "De Instrumentele Ontwikkeling: Analytische Chemie," in *De Geschiedenis van de Scheikunde in Nederland 3*, Homburg and De Palm, 87-103.

[510] L.J. Brasser et al., "Concise Presentation of SO_2-concentration Data Collected from a Regional Network of Volumetric SO_2 Samples," *Atmospheric Environment* 5 (1971): 505-509; Clarenburg,

instrumental techniques such as mass spectrometry in order to identify toxic organic compounds.[511] Jan Willem Copius Peereboom, also interested in toxicology, employed techniques that were used by biologists, such as light microscopy.[512] His research interests were diverse, but often anticipated topical issues that later became a public environmental concern, for instance on heavy metals in soils.[513] The activities of Clarenburg, Copius Peereboom, Dijkstra, Sundström and Hutzinger illustrate the diversity in research interests and chemical practices of the Dutch environmental chemists.

The Dutch environmental chemistry community grew quickly during the 1970s.[514] This is also witnessed by the number of university chairs in environmental chemistry. In 1965 approximately 1.6 % of the chemistry chairs were occupied by environmental chemists, which increased to 4.1 % in 1980.[515] MC had 287 members in 1971: 4.2 % of all KNCV members.[516] This ratio tripled to 13.1 % in 1993; even in a period of overall growth of the KNCV (cf., Figures 4.2 and 4.3). The growth of environmental chemistry thus roughly parallels the trend

---

"Penalization of the Environment Due to Stench," 333-351. This public concern was at the fore within MC: KNCV, "KNCV-Jaarverslag 1971," *Chemisch Weekblad* July 21 (1972): 8-23.

511 G. Sundström and O. Hutzinger, "Environmental Chemistry of Flame Retardants V^{x}. The Composition of Bromkal ® 70-5 DE-A Pentabromodiphenyl Ether Preparation," *Chemosphere* 5 (1976): 187-190; V. Zitko and O. Hutzinger, "Uptake of Chloro and Bromobiphenyls, Hexachloro and Hexabromobenzene by Fish," *Bulletin of Environmental Contamination and Toxicology* 16 (1976): 665-673.

512 See for example: Jenny H.J. Copius Peereboom-Stegema et al., "Influence of Chronic Cd Intoxication on the Alkaline Phosphatase Activity of Liver and Kidney: Biochemical, Histochemical and Histological Investigations," *Toxicology* 14 (1979): 67-80; J.W. Copius Peereboom, "General Aspects of Trace Elements and Health," *The Science of the Total Environment* 42 (1985): 1-27. Jan Willem Copius Peereboom was also director of the Institute for Environmental Problems (Free University, Amsterdam) and one of the first board members of MC.

513 J.H.J. Copius Peereboom and J.W. Copius Peereboom, "The Intake of Cadmium in the Kempen, an Area in the South of the Netherlands," *Ecotoxicology and Environmental Safety* 18 (1989): 93-108. See also: Harry Perrée, "Jan Willem Copius Peereboom: 'Ik Geloof in de Industrie, maar ze Moet zich Netjes Gedragen'," *Natuur en Milieu* 21 (1997): 14-15.

514 See for the growing demand for environmental chemistry research: Ministerie van Volksgezondheid & Milieuhygiëne, "Urgentienota Milieuhygiëne," 1-2; Dinkelman, *Verzuring en Broeikaseffect*, 80-84.

515 These data are derived from Homburg and Palm. It depends on demarcation what counts as environmental chemistry. See: Ernst Homburg and Lodewijk Palm, "Grenzen aan de Groei – Groei aan de Grenzen: Enkele Ontwikkelingslijnen van de Na-oorlogse Chemie," in *De Geschiedenis van de Scheikunde in Nederland 3*, Homburg and De Palm, 3-18. See for the growth in environmental chemistry education: Werkgroep Aspekten van Milieuproblematiek in de Chemische Opleiding, *De Milieuchemie in het Wetenschappelijk Onderwijs* (Den Haag: KNCV, 1976), 1-68; J.W. Copius Peereboom and K. Bouwer, "Environmental Science 'Milieukunde' in the Netherlands: a Review," *The Science of the Total Environment* 129 (1993): 157-170.

516 KNCV, "KNCV-Jaarverslag 1971," 9 and 20.

in public environmental concerns, with a sharp increase since the mid-1980s, partly due to new concerns such as ozone depletion.[517]

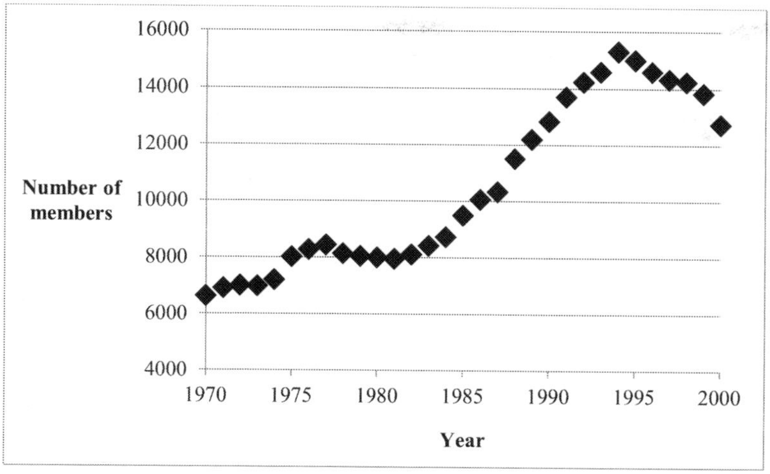

Figure 4.2: Number of KNCV members.[518]

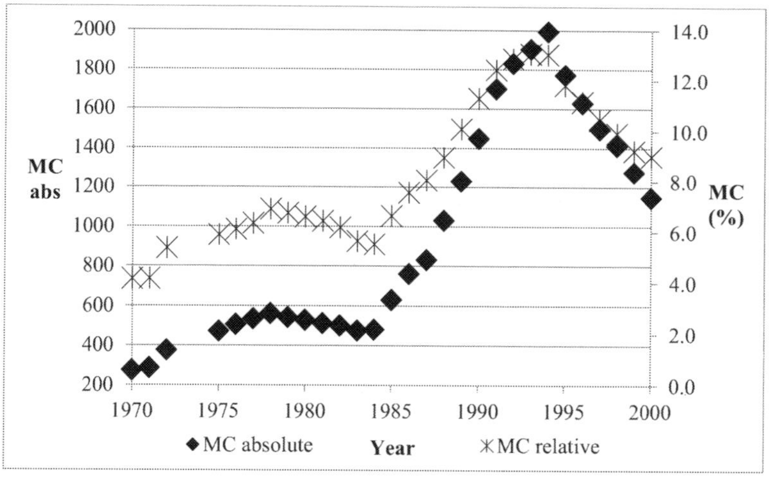

Figure 4.3: The relative (MC (%)) and absolute growth (MC abs) in the number of MC members within the KNCV.[519]

---

[517] See for ozone layer concerns: E.H. Adema et al., "The Heterogeneous Formation of N_2O in Air Containing NO_2, O_3 and NH_3," *Journal of Atmospheric Chemistry* 11 (1990): 255-269.
[518] Data were received from the KNCV administration office.
[519] Ibid. The membership figures of 1973 and 1974 are missing.

The increased attention for the environment within the KNCV continued during the early 1990s.[520] But there were at the same time worries inside the KNCV and the MC that much public unrest on environmental issues was caused by ignorance of chemical matters. The MC even proposed to publish a column in the KNCV-monthly *Chemisch Magazine* about errors made by journalists in the daily press about chemistry and the environment.[521] But the MC also organized symposia that mirrored the complexity of understanding the environment in chemical terms. For example, at the KNCV summer congress of 1990 a symposium was organized on the pathways of chlorinated aromatics in the environment.[522] And in 1995 the MC organized then a symposium about biofuels, lifecycle analysis and catalysis.[523] The Division also collaborated with other divisions and interest groups within the KNCV, such as the Geochemical Circle [Geochemische Kring], the Division of Analytical Chemistry [Sectie Analytische Chemie] and also with the Division of Environmental Toxicology [Sectie Milieutoxicologie] that was part of the Dutch Society for Toxicology [Nederlandse Vereniging voor Toxicologie].[524] This diversity of interests and topics had implications for the scientific identity of Dutch environmental chemistry. Therefore a member of the board of the MC argued at a meeting in 1995:[525]

> (…) that we [the board of the Division] do not exactly know what environmental chemistry is, which activities we want to deploy, (…)

Apparently, after decades of Dutch environmental chemistry, the identity of this subfield of the chemistry discipline was still ambiguous for some members of the MC. Environmental chemistry was not a cognitively unified subfield within the chemical sciences. Dutch environmental chemists, similar to their peers from the USA (Chapter 2) and UK (Chapter 3), were mainly focused on chemical analysis instead of the design of new substances, which contrasts with the "operational realism" concept for chemists in general.[526]

---

[520] See for example: Minutes Executive Committee Meeting of MC, January 24, 1990.
[521] Minutes Executive Committee Meeting of MC, September 27, 1990.
[522] MC, *1990 Annual Report*.
[523] J.J.M. Mulderink, "KNCV-Zomercongres 1995: Natuur, Bron voor Chemie," *Chemisch Weekblad* 91 (1995): 5. See also: Minutes Executive Committee Meeting of MC, April 4, 1995.
[524] KNCV, "KNCV Jaarverslag 1990," *Chemisch Weekblad* 87 (1991): 196; Minutes Executive Committee Meeting of MC, September 9, 1993; MC, *1990 Annual Report*. The MC merged in 2007 with the Division of Environmental Toxicology. See: MC, *2006 Annual Report*.
[525] [(…) dat we niet precies weten wat we onder milieuchemie verstaan, welke activiteiten we willen ontplooien, (…)] As in: Minutes Executive Committee Meeting of MC, April 4, 1995.
[526] Bernadette Bensaude-Vincent and Jonathan Simon, *Chemistry: The Impure Science* (London: Imperial College Press, 2008), 6. See also Chapter 1.

The demarcation of Dutch environmental chemistry must be found in the environment itself, which inspired a whole array of chemical practices and ideas that ranged from applied research to fundamental research interests, often spurred by topical environmental issues.

4.3 The Dutch Approach: Pollution Prevention Policies and Sustainable Development

The USA and the UK introduced pollution prevention policies from the late 1980s onwards.[527] This ran parallel to a shift in the environmental policies of the OECD (Organisation for Economic Co-operation and Development). At their 1985 meeting the Ministers of the Environment of the OECD countries decided to focus on 'Integrated Pollution Prevention and Control'.[528] So internationally, environmental policies tended to shift from command-and-control towards pollution prevention.[529] The Netherlands took part in that process.

As mentioned above, during the late 1980s a 'constructive dialogue' had started between Dutch government and the chemical industry. New government-industry relations developed that culminated into covenants rather than unilateral policy measures. In 1985, for instance, the government started the so-called "Koolwaterstoffen 2000" (KWS 2000) project that aimed at decreasing the relatively high industrial and household emissions of volatile organic compounds (VOC).[530] In his analysis of the policy trajectory of KWS 2000, the Dutch environmental sociologist Arthur Mol showed that representatives of industries, regional and national governments and consumer organizations participated in this consensus seeking project. This led in 1989 to one of the first Dutch environmental covenants, which aimed at a 50 % reduction of VOC emissions by 2000 as compared to 1980.[531]

During the early 1990s, the broadly conceived participatory style of policy making culminated in the establishment of many new environmental covenants between the

---

[527] See Chapters 2 and 3.

[528] Bill L. Long, *International Environmental Issues and the OECD 1950-2000* (Paris: OECD, 2000). Long is the former OECD Director for the Environment.

[529] Command-and-control policy is also called end-of-pipe policy.

[530] See for a comprehensive overview: Mol, *The Refinement of Production*, 153-162. According to the Dutch Government, the USA environmental quality standards for VOCs were exceeded at several places in the Netherlands leading to discussions.

[531] Ed H.T.M. Nijpels, "Rijksbegroting voor het Jaar 1989," letter to the Dutch House of Representatives, April 10, 1989. This covenant was even in advance of the VOC protocol within the *LRTAP* regime, see: Chasek, Downie and Brown, *Global Environmental*, 125.

government and the chemical industry.[532] This style of policy making was time consuming due to the involvement of many stakeholders. But the execution of unilateral measures was time consuming too for the Dutch government, especially when there was a lack of support by stakeholders. Moreover, the government had experienced that unilateral regulatory measures were expensive in their implementation, and also inflexible.[533] The relationship among Dutch government bodies and the chemical industry, generally speaking, developed in a broadly construed participatory style of consensus seeking policies, with the input of scientists, NGOs and industry organizations.

That change in the style of policy making took place in a time of increasing interest in the environment by chemists (cf., Figure 4.3), and of increasing public concerns regarding pollution in general and emissions of CFCs and CO_2 in particular. Members of the Dutch House of Representatives also paid attention to these issues and to the pollution prevention approaches that were emerging.[534] With respect to the Netherlands, it is hard to trace the idea of pollution prevention to one single event, but it was addressed by some participants already during the 1970s. For example, in 1979 Ad Lansink, a chemist and Christian Democrat MP, tabled a motion in the House of Representatives.[535] He advocated a hierarchy in handling waste, with pollution prevention as the option of first choice. His motion received political recognition, as it passed the Dutch House of Representatives. Also the *LRTAP*, which was signed in 1979 and entered into force in 1983, mentioned pollution prevention as a fundamental policy principle.[536]

According to the Netherlands Organization for Technology Assessment [Nederlandse Organisatie voor Technologisch Aspectenonderzoek] (NOTA), government expenditures on

---

[532] Mol, *The Refinement of Production*, 363. In 1993, for instance, a broadly construed environmental covenant was signed between the Government and more than 100 chemical industries. Owing to the involvement of branch organizations, which functioned as intermediary bodies between the Government and individual firms, Mol characterizes the Dutch policy style as neo-corporatism. See also: VNCI, *Jaarverslag 1993* (Den Haag: VNCI, 1994), 9; Koos Schwartz, "Milieuconvenant Stemt tot Tevredenheid," *Trouw*, January 18, 1996, accessed May 22, 2022, https://www.trouw.nl/home/milieuconvenant-stemt-tot-tevredenheid~a630f24a/.

[533] Mol, *The Refinement of Production*, 162. For comprehensive reviews on the consensus seeking political culture of the Netherlands, see: Dennis Bos, Maurits Ebben and Henk te Velde, ed. *Harmonie in Holland: Het Poldermodel van 1500 tot Nu* (Amsterdam: Uitgeverij Bert Bakker, 2011).

[534] See for example: Tweede Kamer Commissie voor Milieubeheer, "Klimaatverandering door CO_2 en Andere Sporegassen: Lijst van Vragen," October 6, 1987.

[535] J. Cramer, B. de Laat and G. Straten, *Preventie van Afval en Emissies in Nederland* (Den Haag: NOTA, 1991), 61-62. In fact, even the Urgency Report Environmental Health of 1972 addressed pollution prevention policies. Environmental concerns, including prevention policies, captured headlines of newspapers. See for example: "Kritiek Kamer Tijdens Begrotingsdebat: Ginjaar te Langzaam met Milieuwetgeving," *Nieuwe Leidsche Courant*, November 1, 1979, 3.

[536] See Chapters 2 and 3.

pollution prevention policies gradually grew during the 1980s.[537] In 1988, for instance, the so-called Project Industrial Successes with Waste Prevention [Project Industriële Successen Met Afvalpreventie] (PRISMA) started that was financially supported by the Ministries of Economic Affairs and of VROM.[538] Some industries collaborated with universities in this pollution prevention policy program that was based on the idea of 'self-regulation' by companies.[539] According to NOTA, this new policy approach resulted in benefits for both the companies involved and the environment. And in 1989 a pollution prevention office was established within VROM, which predominantly concerned the prevention of solid wastes.[540]

Triggered by the famous United Nations publication of *Our Common Future* in 1987, the new Minister of VROM, and member of the liberal party, Ed Nijpels, was convinced by the late 1980s that a 'start' should be made with pollution prevention, in close cooperation with important stakeholders such as environmental NGOs and industry.[541] He advocated an integrated policy strategy that culminated in a huge science-based policy program that set out hundreds of measures: the National Environmental Policy Plan [Nationaal Milieubeleidsplan] (NEPP).[542] The idea of pollution prevention became part of NEPP, which was in turn based on *Concerns for Tomorrow: A National Environmental Survey 1985-2010* [*Zorgen voor Morgen: Nationale Milieuverkenning 1985-2010*], a very influential study published by the

---

[537] Hans Dieleman et al., *Kiezen voor Preventie is Winnen: Naar een Preventief Milieubeleid van Bedrijf en Overheid* (Den Haag: NOTA, 1991), 31-32. In 1994, NOTA was renamed the Rathenau Institute: "What We Do," Rathenau Institute, accessed May 22, 2022, https://www.rathenau.nl/en/about-us/what-we-do. See also: Cramer, De Laat and Straten, *Preventie van Afval en Emissies*, 47; Ritzen, "Wetenschapsbudget 1991," 188.

[538] For an overview of PRISMA, see: Dieleman et al., *Kiezen voor Preventie is Winnen*, 13-23; T.J.N.M. de Bruijn, F.H.J.M. Coenen and K.R.D. Lulofs, "Pollution Prevention Projects in the Netherlands," *Journal of Cleaner Production* 4 (1996): 41-53. For other pollution prevention policies of the Government or NGOs, see: Cramer, De Laat and Straten, *Preventie van Afval en Emissies*, 8-15.

[539] Dieleman et al., *Kiezen voor Preventie is Winnen*, 220. The policy approach of self-regulation coexisted with command-and-control policies.

[540] Ibid., 32.

[541] E.H.T.M. Nijpels, "Preventie en Hergebruik van Afvalstoffen," letter to the Dutch House of Representatives, October 10, 1988. Nijpels wrote this letter also on behalf of the Minister of Economic Affairs, Minister of Agriculture and Fishery, and the Minister of Transportation and Water Management. See also: United Nations World Commission on Environment and Development, *Our Common Future* (Oxford: Oxford University Press, 1987); Dinkelman, *Verzuring en Broeikaseffect*, 193-201.

[542] E.H.T.M. Nijpels et al., preface to Tweede Kamer der Staten-Generaal, "Kamerstuk 21137: Nationaal Milieubeleidsplan (NMP)," May 25, 1989. According to Bennett, the Netherlands was one the first countries with a huge comprehensive policy plan in order to protect the environment. Bennett, "The History of the Dutch," 7.

Dutch National Institute for Public Health and the Environment [Rijksinstituut voor Volksgezondheid en Milieu] (RIVM) in 1988.[543]

As a research institute, the RIVM was the scientific advisor to government agencies and ministries on matters concerning public and environmental health.[544] The publishing of the above mentioned science-based study of future trends with respect to the environment and the climate was part of that role. During the preparation of the report the RIVM collaborated with many different government agencies, engineering companies, scientists, ministries and universities. That fitted well in the Dutch consensus seeking civic epistemology.[545]

The RIVM report *Concerns for Tomorrow* discusses environmental issues and climate change, based on the sustainable development concept formulated in *Our Common Future.*[546] Minister Nijpels emphasized in his preface to the report that sustainable development should become the leading objective in new environmental policies.[547] He also pointed to major new policy challenges that were not limited to geographical borders, such as acid rain, global warming and ozone depletion.[548] Besides these issues of geopolitical concern, *Concerns for Tomorrow* highlighted typical Dutch environmental problems such as the huge fossil energy consumption per km^2 and bioaccumulation of toxics.[549] These problems were traditionally approached through end-of-pipe sanctions.[550] This style of policy making had its advantages, but its disadvantages too, as 'end-of-pipe sanctions were expensive'.[551] These economic

---

[543] E.H.T.M. Nijpels, preface to *Zorgen voor Morgen*, Langeweg, VII-VIII; The RIVM was established in 1984 as a merger of three government agencies that operated under the authority of the Minister of Health Affairs. For a comprehensive history of RIVM, see: Van Zon, *Tachtig Jaar*.
[544] Van Zon, *Tachtig Jaar*, 405.
[545] The RIVM collaborated for instance with the Institute for Environmental Problems, Directorate-General Environmental Protection (VROM) and the environmental scientist Lucas Reijnders (Foundation for Nature and the Environment). See: R.B.J.C. van Noort, preface to *Zorgen voor Morgen*, Langeweg, IX-XI. Van Noort was the Director-General of the RIVM. See for the civic epistemology concept Chapter 1 and see: Sheila Jasanoff, *Designs on Nature: Science and Democracy in Europe and the United States* (Princeton, NJ: Princeton University Press, 2005), 247-271.
[546] Langeweg, *Zorgen voor Morgen*, 3; United Nations World Commission on Environment and Development, *Our Common Future*.
[547] Nijpels, preface to *Zorgen voor Morgen*, Langeweg, VII-VIII.
[548] Ibid. Global warming is a popular term for climate change. See also: Anthony N. Penna, *The Human Footprint: A Global Environmental History* (Chichester: Wiley-Blackwell, 2010), 287.
[549] Traffic caused emissions of toxic compounds too. In the 1980s the Netherlands had the highest number of cars per square kilometer as compared to many other countries. See: Langeweg, *Zorgen voor Morgen*, 42-49.
[550] The "Indicatief Meerjaren Programma Afvalstoffen Chemisch Afval 1985-1989 (IMP-CA)" policy that was launched in 1984 also featured some pollution prevention ideas. See: VNCI, *Jaarverslag 1984*, 23; Gerard Keijzers, "The Evolution of Dutch Environmental Policy: The Changing Ecological Arena from 1970-2000 and Beyond," *Journal of Cleaner Production* 8 (2000): 179-200.
[551] F. Langeweg, summary to *Zorgen voor Morgen*, by Langeweg, XIII-XXVIII.

disadvantages were recognized in the RIVM report and therefore pollution prevention policies should get a larger role, next to the traditional policy instruments.[552]

Based on *Concerns for Tomorrow*, the Ministry of VROM issued the first NEPP in 1989 and NEPP-plus in 1990.[553] Graham Bennett, heading the Institute for European Environmental Policy (Arnhem, the Netherlands), reviewed the development of NEPP, and highlighted that NEPP was aiming for sustainable development.[554] Bennett contended that NEPP was based on eight principles:[555]

> (…) the "standstill" principle to ensure that environmental quality standards does not degrade any further; pollution abatement at the source; the "polluter pays" principle; prevention of unnecessary pollution; application of the best practicable means of pollution abatement; application of the principles for isolating, controlling, and monitoring wastes; source-oriented controls based on effect-oriented quality standards; and internalization of environmental control into the activities of the target groups (those that will bear the brunt of the costs of implementing environmental control measures).

The focus on pollution prevention, in combination with environmental control and sustainable development, would shape Dutch environmental policies during the next decade. The NEPP provisions formed the basis for environmental quality standards – for instance on emissions of ammonia (NH_3), SO_2 and NO_x – that were strict in comparison to those of other countries of the European Community.[556] Partly influenced by international environmental disasters, public concerns were one of the political reasons for this environmental policy.[557] According

---

552 Ibid., XXI. See also: Dinkelman, *Verzuring en Broeikaseffect*, 125. The VNCI recognized this pollution prevention interest too, with 'waste prevention is high on the political agenda'. See: VNCI, *Jaarverslag 1988*, 16.

553 E.H.T.M. Nijpels, "Nationaal Milieubeleidsplan (NMP)," letter to the Dutch House of Representatives, May 25, 1989. Nijpels wrote this letter also on behalf of the Minister of Economic Affairs, Minister of Agriculture and Fishery, and the Minister of Transportation and Water Management. See also: M.E. Minnesma, "Dutch Climate Policy: A Victim of Economic Growth?," *Climate Policy* 3 (2003): 45-56; Bennett, "The History of the Dutch," 7-9 and 31-33.

554 Bennett, "The History of the Dutch," 7-9 and 31-33. See also: Nijpels et al., preface to Tweede Kamer der Staten-Generaal, "Kamerstuk 21137: Nationaal Milieubeleidsplan (NMP)"; Nijpels, preface to *Zorgen voor Morgen*, Langeweg, VII-VIII.

555 Bennett, "The History of the Dutch," 8; Tweede Kamer der Staten-Generaal, "Kamerstuk 21137: Nationaal Milieubeleidsplan (NMP)," 12.

556 Bennett, "The History of the Dutch," 7-9 and 31-33; VNCI, *Jaarverslag 1989*, 13-16; R.E. Selman, preface to VNCI, *Jaarverslag 1990* (Den Haag: VNCI, 1991), 3-4. According to Selman, the OECD had the same opinion on this as the VNCI. See also: Bottelberghs, "Risk Analysis and Safety Policy," 59-84.

557 Bennett, "The History of the Dutch," 7-9 and 31-33; Bottelberghs, "Risk Analysis and Safety Policy," 59-84.

to Bennett, the development of NEPP had a 'long and difficult birth' due to time consuming negotiations among role-based stakeholders with diverging interests and provided, in a period of economic growth and heightened public environmental concerns, the basis of strict environmental quality standards.[558] This illustrates its value to Dutch society, and most notably in the political arena. The policies following from the NEPP were continued during the early 1990s, with growing government expenditures to environmental protection, a focus on pollution prevention, environmental covenants and the launch of the Environmental Protection Act [Wet Milieubeheer] in 1993, which integrally covered all sorts of environmental protection.[559] The RIVM report *Concerns for Tomorrow*, as well as these new environmental policy programs, also formed the basis of a new science policy initiative that was launched during the early 1990s, and that, as will be shown in the next section, would shape the rise of new initiatives and a new scientific 'rationale' for the chemical science under the banner sustainable chemistry.[560]

4.4 The 1990s: a New Chemistry Oriented Network in the Netherlands

In 1990 the KNCV published *The Environment: Why Chemists Care* [*Milieu: Chemici een Zorg*] as a response to *Concerns for Tomorrow* and the NEPP.[561] This report by the KNCV contained an important recommendation for the government and Dutch society: chemists can and should be part and parcel of the road to be taken in order to achieve sustainability.[562] The KNCV thus agreed with the principal ideas behind the NEPPs such as pollution prevention, but also urged government to support fundamental research on a wide range of topics that

---

[558] Bennett, "The History of the Dutch," 7 and 8-9; Bosmans and Van Kessel, *Parlementaire Geschiedenis*, 165-184.
[559] Cramer, De Laat and Straten, *Preventie van Afval en Emissies*, 47; Dieleman et al., *Kiezen voor Preventie is Winnen*, 31-32; Ritzen, "Wetenschapsbudget 1991," 188. See also: "Wet Milieubeheer," *Staatsblad van het Koninkrijk der Nederlanden* 80 (1994): 1-98; Bottelberghs, "Risk Analysis and Safety Policy," 59-84. For details about pollution prevention projects, see: De Bruijn, Coenen and Lulofs, "Pollution Prevention Projects," 41-53. De Bruijn et al. stated that several pollution prevention projects in the Netherlands were comparable to the "Waste Minimization Opportunity Assessment Manual, 1989" of the US Environmental Protection Agency (EPA).
[560] Hessels and Van Lente, "Practical Applications as a Source of Credibility," 218.
[561] Minutes Executive Committee Meeting of MC, March 28, 1990; Taskforce KNCV, *Milieu: Chemici een Zorg*, 1-2 and 20. This taskforce consisted of members representing governments, universities, KNCV, MC, VNCI, RIVM, industries and the national Foundation for Nature and the Environment. See also: Yvonne Huigen, "Milieu, Chemici een Zorg," *Chemisch Weekblad* 87 (1991): 6.
[562] KNCV members requested for this report. See: Taskforce KNCV, *Milieu: Chemici een Zorg*, 2. See also: KNCV, "KNCV Jaarverslag 1990," 196.

could be relevant for the environment.[563] The KNCV addressed in its report also two intertwined concerns: the chemical literacy of the public, and the chemical knowledge needed for environmental control.[564] The ideas put forward were discussed among chemists, and proved to be also influential in science policy.

During the early 1990s the chemical community undertook several actions aiming at the improvement of the public chemical literacy and at the 'conditions' for chemical science to function well, next to the tackling of environmental problems by means of chemical knowledge.[565] For example, the KNCV Platform Chemistry and Environment [KNCV Platform Chemie en Milieu] was launched in 1991 in order to coordinate actions that addressed these objectives.[566] In the same year, both *Chemisch Weekblad* and *Chemisch Magazine* published a special issue on the negative public image of chemistry.[567] These initiatives culminated in a professional code of conduct that was based on *The Environment: Why Chemists Care* and on the results of a survey by the KNCV Working Party Professional Conduct of Chemists [KNCV Werkgroep Beroepscode Chemici]. The code was formulated in 1993 as follows:[568]

> The society [KNCV] supports the scientific integrity and the societal sense of responsibility of its members to chemistry related actions – both in professional actions and outside it – and the consequences of that for nature and society.

The KNCV urged for a proper view and appreciation of the chemical sciences at large by 'the general public'.[569] In 1993, the Foundation Communication Centre Chemistry [Stichting Communicatie Centrum Chemie] (C_3) was established by the KNCV in cooperation with the

---

563 Taskforce KNCV, *Milieu: Chemici een Zorg*, 4 and 16.
564 Ibid.
565 Hessels and Van Lente, "Practical Applications as a Source of Credibility," 218.
566 KNCV, "KNCV Jaarverslag 1991," *Chemisch Weekblad* 88 (1992): 156; KNCV, "KNCV Jaarverslag 1992," *Chemisch Weekblad* 89 (1993): 12. See also: Taskforce KNCV, *Milieu: Chemici een Zorg*, 19.
567 KNCV, "KNCV Jaarverslag 1991," 155. The Editorial Board of *Chemisch Weekblad* was the same as the one of *Chemisch Magazine*. See also: Marion de Boo and Peter van de Beek, "Diepgeworteld Wantrouwen," *NRC*, March 14, 1991, accessed May 22, 2022, https://www.nrc.nl/nieuws/1991/03/14/diepgeworteld-wantrouwen-6960265-a266131. De Boo and Van de Beek discussed the public image concerns of chemists in this widely read Dutch newspaper.
568 [De vereniging bevordert de wetenschappelijke integriteit en het maatschappelijk verantwoordelijkheidsgevoel van haar leden bij het aan de chemie gerelateerde handelen – zowel in de beroepsuitoefening als daarbuiten – en de consequenties daarvan voor natuur en samenleving.] As in: KNCV, "KNCV Jaarverslag 1993," *Chemisch Weekblad* 90 (1994): 7. See also: Taskforce KNCV, *Milieu: Chemici een Zorg*, 13.
569 KNCV, "KNCV Jaarverslag 1993," 7.

VNCI and the Netherlands Association for Science Education [Nederlandse Vereniging voor het Onderwijs in de Natuurwetenschappen].[570] The target groups of C_3 were pupils, students, environmental NGOs as well as the general public, with the aim to improve the dialogue between the chemical community and Dutch society.[571] Both the negative public image of chemistry and the state of the environment clearly were important issues for the Dutch chemical community in those years.

Prevention of environmental problems, instead of end-of-pipe measures, did have an impact on the type of chemical knowledge that would be required to solve environmental problems. A striking example is a special issue of *Chemisch Magazine* from 1991, called "green chemistry" ["groene chemie"]. All articles were on the exploitation of biomass as a renewable feedstock. The authors clearly considered "green chemistry" to be a chemical practice based on renewable, "green", feedstocks.[572] In a comment to the special issue, science journalist Marian van Opstal stated that 'green chemistry is more environmentally friendly than petrochemistry'.[573] This underlined that the use of feedstocks were of central importance for Dutch chemists at that time.[574] That perception of green chemistry in the early 1990s, which was quite different from the way this term was used in the USA, continued during most of the 1990s. It was used in some communications in international journals and, especially, by the Dutch policy network Sustainable Technological Development Chemistry [Duurzame Technologische Ontwikkeling Chemie (DTO-Chemie)] (STD-Chemistry) that was created shortly after the "green" special issue was published.[575]

---

570 Ibid.; VNCI, *Jaarverslag 1993*, 8.

571 KNCV, "KNCV Jaarverslag 1993," 7. The KNCV experienced decreasing numbers of chemistry students too. Nowadays, C_3 mainly focuses on students at secondary schools.

572 A. Fuchs and H. van Bekkum, "Inuline, het Vergeten Zusje van Zetmeel," *Chemisch Magazine* 6/7 (1991): 320-321; J.R. van den Berg, "Groene Chemie ook Interessant voor Shell," *Chemisch Magazine* 6/7 (1991): 322-329; T.J.H.M. van Hutten, "Landbouw en Industrie Moeten Elkaar Leren Verstaan," *Chemisch Magazine* 6/7 (1991): 326-329; J. de Bruyn, "Behoefte aan Grondstoffen op Maat," *Chemisch Magazine* 6/7 (1991): 330-334; H. van Bekkum and A. Fuchs, "De Verwachtingen Zijn Hooggespannen," *Chemisch Magazine* 6/7 (1991): 334-339; J.A. Roels, "Nu Slechts Veelbelovend, Straks Wellicht Realiteit," *Chemisch Magazine* 6/7 (1991): 339-341.

573 Marian van Opstal, "Addertjes onder het Groen," *Chemisch Magazine* 6/7 (1991). The special issue was partly based on the symposium titled Plantaardige Grondstoffen voor de Industrie that was organized by the Netherlands Study Centre for Technology Trends [Stichting Toekomstbeeld der Techniek] on April 25, 1991.

574 See for a recent example of the Dutch focus on the efficient use of preferably renewable feedstocks in chemical processes, including the use of waste as a resource: Tom Keijer, Vincent Bakker and J. Chris Slootweg, "Circular Chemistry to Enable a Circular Economy," *Nature Chemistry* 11 (2019): 190-195.

575 See for example: C. Okkerse and H. van Bekkum, "From Fossil to Green," *Green Chemistry* 1 (1999): 107-114; Leontine A. de Graaf and Peter Kolster, "Industrial Proteins as a Green Alternative for 'Petro' Polymers: Potentials and Limitations," *Macromolecular Symposia* 127 (1998): 51-58.

That network was a "spin-off" of the Dutch interdepartmental research program Sustainable Technological Development [Duurzame Technologische Ontwikkeling (DTO)] (STD) initiated by Dutch government in 1992, for the years 1993 to 1997. The research program was one of the follow-ups to *Our Common Future*. STD-Chemistry emerged out of this new science policy program. It had a great impact on the vocabulary on (green and) sustainable chemistry that emerged in the Netherlands. This took place partly with the help of government agencies, which fitted well to the participatory style of policy making, aiming at consensus among various stakeholders. In this process a programmatic alliance was created between science, state and industry.

STD was executed under the leadership of the science policy experts Leo Jansen and Philip Vergragt, who were both chemists by training.[576] Based on the NEPP of 1989, political initiatives developed that urged for radical technological change aiming at sustainable development.[577] STD was the answer to those pleas and 'sponsored by five government ministries' and by some funding from industry, NGOs and academia.[578] It aimed at the development of new technology with an increase in '"environmental efficiency" by a factor 10 to 50 in the next 50 years' in order to anticipate, for instance, the 'depletion of resources' due to the growing world population.[579] According to the science policy analyst Barend van der Meulen, the STD program followed the so-called foresight principle:[580]

> The quintessence of the idea of 'foresight' is that future studies should not try to forecast. Instead, by presenting visions of the future, the studies should stretch the time horizon of science and technology policy strategies and challenge actors to take part in developing the future. The outcome is a process of exploring possible futures rather than the specific predictions of how the future will be. In technology foresight the process is focused on identification of (critical, generic) technologies that are likely to have a large impact on economic, ecological and social developments.

---

576 Philip J. Vergragt and Leo Jansen, "Sustainable Technological Development: The Making of a Dutch Long-term Oriented Technology Programme," *Project Appraisal* 8 (1993): 134-140.
577 See for example: J.E. Andriessen, "Technologie en Milieu," letter of the Minister of Economic Affairs and Minister of VROM to the members of the Dutch House of Representatives, April 1, 1992.
578 Vergragt and Jansen, "Sustainable Technological Development," 134.
579 Ibid., 136; See also: Jan J.M. Mulderink, preface to DTO-Chemie, *Duurzame Technologische Ontwikkeling Chemie: Vijf Technologie Ontwikkelingsvelden; Onderzoek op Maat* (Den Haag: Willems & Van den Wildenberg, 1997); Andriessen, "Technologie en Milieu". Major companies like DSM, Akzo-Nobel and Shell provided additional funding.
580 Barend van der Meulen, "The Impact of Foresight on Environmental Science and Technology Policy in the Netherlands," *Futures* 31 (1999): 8. See also: Ritzen, "Wetenschapsbudget 1991," 13. See Chapter 3 for "foresight" in the UK.

STD had two target groups. The first were the general public and societal opinion leaders, 'who should regain trust in technology as a contribution to sustainability'.[581] It was believed that aiming at sustainability required recognition by the public, politicians included. The second target group were scientists and technologists, who, according to Jansen, would be willing to cooperate in STD because 'industry and science and technology were suffering from a bad image' and could improve their image by doing something sustainable.[582]

STD was not a new "laboratory", but 'a practical experiment to influence existing networks or as an attempt to create new ones'.[583] As a result of this government initiative, approximately 600 technologists and scientists from industry, government and academia became, more or less, involved in this new science policy program.[584] An additional 2000 technologists, scientists, policymakers and users of knowledge, attended congresses, and symposia, in order to balance vested interests and strong beliefs concerning the shaping of the future by scientific knowledge. This illustrates the Dutch decentralized consensus seeking style of strategic science policy making of the 1990s.

This style 'depends on social control and credibility pressure' in order to create a collective strategy rather than top-down approaches by a government.[585] Van der Meulen and his colleague Arie Rip therefore concluded:[586]

> The national research system of the Netherlands exemplifies general characteristics of modern research systems, with an intermediary level between the state and research performance organisations, and with increasing pressure for relevance of science. Specific for the Netherlands is the [high] density of institutions in the intermediary level and the institutional competence in (heterogeneous) aggregation processes for agenda building, and consociational implementation.

---

581 Vergragt and Jansen, "Sustainable Technological Development," 137. See also: Van der Meulen, "The Impact of Foresight," 15; C. Okkerse and H. van Bekkum, "Vernieuwbare Grondstoffen voor de Chemische Industrie," in DTO-Chemie, *Duurzaamheid en Chemie: Een Bundel Essays* (Delft: DTO, 1996), 42. According to Okkerse and Van Bekkum, chemical education and research largely ignored the urge for a sustainability switch, while it could improve the public image of chemistry.

582 Leo Jansen, "The Challenge of Sustainable Development," *Journal of Cleaner Production* 11 (2003): 239. With respect to the VNCI, Mulderink argued the same: Jan J.M. Mulderink, interview, August 30, 2011.

583 Vergragt and Jansen, "Sustainable Technological Development," 135.

584 Jansen, "The Challenge," 231-245. Mulderink argued that dozens of scientists were involved at the periphery: Mulderink, interview, August 30, 2011.

585 Barend van der Meulen and Arie Rip, "Mediation in the Dutch Science System," *Research Policy* 27 (1998): 758. See also: Van der Meulen, "The Impact of Foresight," 9.

586 Van der Meulen and Rip, "Mediation," 757.

Several sectors were identified within STD, and among those sectors were the chemical sciences: STD-Chemistry.[587] As chair of STD, Jansen approached the former Akzo R&D leader Jan Mulderink for heading STD-Chemistry. Mulderink was interested in sustainability and had, as industrial leader and former KNCV President, an excellent network in the chemical arena.[588] As KNCV President he had organized in 1995, in cooperation with Jansen, a symposium on sustainable development and natural feedstocks in chemistry at the KNCV summer congress.[589] Jansen, a former member of the Dutch House of Representatives and trained as a chemist, also lectured at that congress on the STD initiative.[590]

Five technological research areas were identified within STD-Chemistry for the 'translation of environmental policy objectives into research strategies and research projects' that often concerned the efficient use of (preferably renewable) feedstocks with economically viable perspectives.[591] They included: (1) whole crop utilization [Geïntegreerde plantconversie] (GPC); (2) conversion of hydrocarbons [Koolwaterstoffenconversie] (C_1Chemistry); (3) photovoltaic solar energy [Fotovoltaïsche zonne-energie] (PZC); (4) fine chemicals process technology [Fijnchemische procestechnologie] (PFC); and (5) fiber-reinforced composite materials [Vezelversterkte composiet-materialen] (VVC).[592] Both GPC and C_1Chemistry concerned the utilization of "green", renewable, feedstocks (biomass). PZC would provide 'dematerialized energy', whereas PFC should improve the efficiency of chemical processes in order to prevent pollution.[593] VVC was considered important, because the chemical industry produced significant amounts of construction materials.

After the identification of these R&D areas as a result of the first round of the STD program, it was time for action in the laboratory, in order to show real impact on the basis of so-called "illustration projects". To this end, the Foundation for Sustainable Chemistry

---

587 Anonymous, "DTO Chemie," in *Duurzame Technologische Ontwikkeling Chemie*, DTO-Chemie; Jan J.M. Mulderink, "Sustainability in the Chemical Industry by 2050: The Outlook for a Sustainable Chemical Industry in 2050," *Studies in Environmental Science* 72 (1998): 909-913.
588 In 1994 and 1996, Mulderink was Vice President of the KNCV; in 1995 he chaired the KNCV. He was also one of the founders of the KNCV Division Chemistry and Management. In 2003, he became honorary member of the KNCV.
589 Mulderink, interview, August 30, 2011; Anonymous, "KNCV-Zomercongres 1995: Natuur, Bron voor Chemie," *Chemisch Weekblad* 91 (1995): 5; Anonymous, "Jaaroverzicht 1995," *Chemisch Weekblad* 91 (1995): 3.
590 Parlementair Documentatie Centrum, "Dr. J.L.A. Jansen," Leiden University, accessed May 22, 2022, http://www.parlementairdocumentatiecentrum.nl/id/vg09ller3mxc.
591 Van der Meulen, "The Impact of Foresight," 21-22. See also: DTO-Chemie, *Duurzaamheid en Chemie*; Gerda Sulmann, "Duurzame Chemie Vereist een Sprong Vanúit de Toekomst," *Chemisch Weekblad* 91 (1995): 1.
592 Mulderink, preface to DTO-Chemie, *Duurzame Technologische Ontwikkeling Chemie*.
593 Ibid.

Development [Stichting Duurzame Chemie Ontwikkeling] (DCO) was founded in 1997, as a follow-up to STD-Chemistry.[594] Mulderink continued his leading role by chairing DCO. Just like STD-Chemistry, DCO was not running a concrete laboratory where scientific puzzles were solved, but formed an actor-network instead. The name of this foundation contained the term "sustainable chemistry" ["duurzame chemie"], which was chosen on pragmatic grounds. In the first place, because (in Dutch) the name had resemblances with STD-Chemistry. By using the term sustainable chemistry it was made clear that the new foundation was part of the larger science policy initiative. A second reason was that the term was considered a convenient umbrella term for chemical activities that contributed to sustainable development.[595]

DCO organized several meetings with stakeholders from government, industry and academia, mainly based on the R&D areas identified during the STD-Chemistry program. It was able to pay its day-to-day operations by the financial support from some chemical companies and the VNCI, and to a lesser extent also from government.[596] DCO operated as a broker for R&D grants. This explains much of the interest of academia and industry in DCO. Since 1950, the Dutch Research Council [Nederlandse Organisatie voor Wetenschappelijk Onderzoek] (NWO) was the main Dutch public research funding agency.[597] In 2003, there were plans that DCO should merge with NWO, but these plans did not get enough support.[598] In 2004 DCO submitted two research proposals, amounting to € 30 million each, to NWO. One proposal concerned photovoltaic cells, and the other one the conversion of biomass. If NWO would award these grants to DCO, some companies and academia were prepared to contribute with additional R&D funding too. Unfortunately, both research proposals were rejected in the NWO funding competition. Around that time, the interest by companies and academia in DCO faded too, as did the financial support by government. This was the end of

---

594 See for example: DCO, *Jaarverslag 2003* (Wageningen: DCO, 2004); Tweede Kamer der Staten-Generaal, *Milieuprogramma 2001-2004: Vergaderjaar 2000-2001* ('s-Gravenhage: SDU Uitgevers, 2000), 159.

595 Mulderink, interview, August 30, 2011. Dutch environmental NGOs used this term too: Jan W. Biekart, Anton Nigten and Dick Stoppelenburg, ed., *Duurzame Chemie: Werkstrategieën voor Duurzaamheid in de Chemische Industrie met een Toetsing aan de Huidige Praktijk* (Utrecht: Stichting Natuur en Milieu, Vereniging Milieudefensie and Waddenvereniging, 1997).

596 VNCI, *Jaarverslag 1997* (Den Haag: VNCI, 1998), 13; DCO, *Jaarverslag 2003*; Mulderink, interview, August 30, 2011. See also: Lars Coenen, *DTO Illustratieprocessen, Wordt Vervolgd...?: Onderzoek naar Vervolgtrajecten op de Illustratieprocessen van het Interdepartementaal Onderzoekprogramma Duurzame Technologische Ontwikkeling (1993-1997)* (Eindhoven: TU/e, 2000), 86-89.

597 "Wat NWO Doet," NWO, accessed May 22, 2022, http://www.nwo.nl/over-nwo/missie+en+visie.

598 Mulderink, interview, August 30, 2011; DCO, *Jaarverslag 2003*.

DCO. What remained was the use of the term sustainable chemistry by NWO in funding programs, and in science communication.[599]

The STD-Chemistry and DCO initiatives operated in an industrial-academic network that stood largely outside the MC. Quite different as compared to the USA, the MC Division within the KNCV was not in the lead with respect to the emergence of new concepts such as green chemistry or sustainable chemistry. Dick Sijm, who chaired MC in 1996-1998, stated for instance in 1999 that environmental chemistry should not only concern the pathways, analysis and toxicological effects of substances and particles in the environment, but also take into account the 'design phase':[600]

> By taking into account environmentally relevant properties in the 'design' phase, a lot of problems could be prevented by taking certain substances not in production. This fits in the pursuit of sustainable chemistry ("green chemistry").

Sijm stressed the importance of pollution prevention, but also realized that environmental chemists focused on waste releases to the environment due to pollution caused by chemical processes. With "sustainable chemistry" and "green chemistry", he introduced two terms that were not reported before within MC. Sijm related green chemistry to the avoidance of impact of chemical pollution on the environment. He was familiar with international journals on environmental chemistry, but when interviewed by the author in 2011, Sijm told that he was not aware of the green chemistry and benign-by-design chemistry philosophy of Paul Anastas, or of related activities within the ACS Division of Environmental Chemistry and the EPA. He only became aware of those terms and ideas, he said in 2011, via a colleague from industry.[601] In Sijm's view "sustainable chemistry" as implied, but not mentioned, by the report *Concerns for Tomorrow* was more holistic than "green chemistry". But, just like several other members

---

[599] See for example: NWO Chemische Wetenschappen, *Jaarboek 2005* (Den Haag: NWO, 2005), 10. See also: Ton van Helvoort and Ernst Homburg, *Een Eeuw Chemische Technologie in Nederland* (Delft: Stichting Hoogewerff-Fonds, 2021), 187-215. (See herein: Chapter 10 'Op Weg naar een Duurzame Wereld: Uitdagingen voor de Procestechnologie.)

[600] Dick Sijm, "Nieuwe Uitdagingen in de Milieuchemie," in Sectie Milieuchemie, *Jaarboek 1998: Sectie Milieuchemie* (Den Haag: KNCV, 1999), 98-102; Dick Sijm, interview, August 24, 2011. [Door rekening te houden met milieurelevante eigenschappen in de "design" fase kunnen veel problemen worden voorkomen door het niet in productie nemen van bepaalde stoffen. Dit past in het streven naar duurzame chemie ("green chemistry").] As in: Sijm, "Nieuwe Uitdagingen in de Milieuchemie," 102.

[601] Sijm, interview, August 24, 2011.

of the MC, Sijm was not familiar with the major Dutch sustainable chemistry initiative DCO, that has been discussed above.[602]

The emergence of the terms green chemistry and sustainable chemistry in the Netherlands thus was, at least partly, a response to environmental concerns by the public, and to the negative image of chemistry resulted from that. The term green chemistry had a more limited meaning in the Netherlands, as compared to the USA. It was mainly associated with renewable feedstocks from nature.[603] The term sustainable chemistry, by contrast, had a far wider scope, and included, for instance, photovoltaic, fine chemicals, and special polymers. It was also the term that got a far wider circulations as a result of the STD and DCO funding schemes. Inspired by key-actors such as Jansen, Vergragt and Mulderink, chemists were willing to participate in STD-Chemistry and in DCO as these initiatives were potentially good for the environment, and could improve the public image of chemistry at the same time.[604] Although these two initiatives were in line with the views of KNCV on chemistry and society, and with the aims of the MC, both the KNCV and the MC were only at the periphery in the actor-networks of STD-Chemistry and DCO.

4.5 Discussion: the Dutch Road to Green and Sustainable Chemistry

As demonstrated above social concerns regarding the environment increasingly received a political response in the Netherlands, as witnessed by new legislations in the 1970s, and by increasing membership numbers of environmental NGOs. Compared to other countries, the extent of this "democratization of environmental concerns" and its political impact were large. In the late 1980s, for instance, the environmental quality standards were more strict than in most other countries, and the Netherlands also played a leading role in the transnational *LRTAP* regime. The influential report *Concerns for Tomorrow* and the environmental covenants from the late 1980s onwards, illustrate that the command-and-control policy style shifted to consensus seeking approaches between diverse stakeholders.[605] Diverging interests

---

602 Ibid.

603 See for example: Wetenschappelijke en Technologische Commissie voor de Biobased Economy, *Naar Groene Chemie en Groene Materialen: Kennis- en Innovatieagenda voor de Biobased Economy* (Den Haag: WTC, 2011), 11-13. This advisory committee for the Dutch Government was established in 2010 and consisted of members from government, industries and academia. Though, in 2011 green chemistry was seen as something quite new by this committee. The overall aim in this report was the transition of a petrochemistry to a green chemistry with a strong role for industry.

604 Vergragt and Jansen, "Sustainable Technological Development," 134-140; Jansen, "The Challenge," 231-245.

605 Jasanoff, *Designs on Nature*, 259.

of chemical industry and NGOs were balanced by an active government that supported a broad participatory style of policy making in the Dutch consensus seeking civic epistemology.[606]

These dynamics and mechanisms of Dutch society were also visible in science policy making, and in the actions of professional organizations such as KNCV. The chemical community responded to social concerns about the state of the environment by the introduction of new chemical subfields, such as environmental chemistry, which peaked greatly in the early 1990s (cf., Figure 4.3). In the course of the 1990s a demand for new applied chemical and technological knowledge was the result of the widely felt need of pollution prevention and sustainable development. Those new environmental policy objectives thus shaped a new scientific 'rationale' for chemists, as witnessed by STD-Chemistry.[607] In that policy program, many members from academia, industry, NGOs and government agencies were involved, which illustrated the broad and decentralized Dutch style of policy making. The STD-Chemistry program helped to make the scientific efforts more coherent, as fields of study were identified on the basis of economic and environmental grounds.

Different from the situation in the US, the chemical society KNCV did not have a leading role. It was not central in STD-Chemistry, which was mainly funded by Dutch government. Though similar to the case of the Royal Society of Chemistry (RSC) in the UK, the KNCV did recognize important policy objectives of this science policy program such as pollution prevention and the improvement of the negative image of chemistry.[608] Through those funding schemes, STD-Chemistry received also recognition from chemists. Moreover, the Dutch chemical sciences were in need of a greater public recognition. According to Jansen and Vergragt, this need also played a role for chemists to join STD-Chemistry, so that they could show that they were aiming for sustainability.

Also KNCV tried hard to improve the public image of chemistry. This led to actions such as the establishment of C_3.[609] Jansen also got a stage at the KNCV summer congress of 1995, to discuss his STD activities.[610] Clearly, the combined idea of sustainable development and chemical knowledge increasingly started to play a central role.

---

[606] Jasanoff, *Designs on Nature*, 259.
[607] See Chapter 1. See also: Hessels and Van Lente, "Practical Applications as a Source of Credibility," 218.
[608] See Chapter 3 for the response of the RSC on Technology Foresight.
[609] VNCI, *Jaarverslag 1993*, 8; KNCV, "KNCV Jaarverslag 1993," 7.
[610] Anonymous, "KNCV-Zomercongres 1995," 5.

As a result, a new vocabulary emerged among chemists aiming at sustainable development and at an improved public image: "sustainable chemistry". This is illustrated by the choice of the name DCO for the foundation promoting these aims. The term fitted well to the broader policy objective of a sustainable development. Applying the idea of the 'science-society contract' of Laurens Hessels and Harro van Lente, the chemical sciences thus tried to change its 'identity' as becoming a sustainable chemistry due to a changing 'rationale', in a context of dialogue among many stakeholders, ranging from users, and producers, of chemical knowledge.[611]

The terms green chemistry and sustainable chemistry coexisted in the Netherlands since the 1990s, but the term sustainable chemistry was more frequently used. The term green chemistry was mainly used, as mentioned above, in a context of renewable feedstocks, which also became part of the STD-Chemistry program. The scientific content of green chemistry and sustainable chemistry have not yet been discussed in great detail though. This will be the topic of the next chapter.

---

611 Hessels and Van Lente, "Practical Applications as a Source of Credibility," 215-240.

Chapter 5: Intellectual Origins of Green Chemistry: Diverging Views on Scope and Content

The previous chapters describe the impact of contextual factors on the growth and emergence of environmental and green chemistry in three different countries. It was argued that the increase of public environmental concerns and environmental legislation paralleled the growth of the membership of sections on environmental chemistry in the three chemical societies investigated. The research topics chosen by environmental chemists were often closely related to topical public concerns. Contextual drivers thus crossed the boundaries of chemical laboratories by impacting the research agendas. The "externalist" histories in the previous chapters also show that chemical societies held the opinion that chemistry was (unjustly) blamed by the public as a cause of environmental pollution, which they saw as the result of a lack of public chemical literacy.[612] Chemical societies had to deal with these aspects, which contributed to the emergence and institutionalization of green chemistry in the course of the 1990s.[613] By contrast, the present more "internalist" oriented chapter examines the nature and conceptual identity of green chemistry in order to understand its growth, disciplinary identity and scientific recognition.[614]

In an earlier publication I have shown that the popularity in using the term "green chemistry" started almost out of the blue in the mid-1990s.[615] As announced in Chapter 1, the present chapter pays special attention to the emergence of the foundations and epistemic components of green chemistry in the 1990s and the early 2000s, i.e., the period characterized by a quite sudden and huge growth of the field. These aspects shaped the internal scientific development of green chemistry. In this chapter I analyze the development of (new) concepts introduced by influential green chemists during this period. I also give some examples of

---

[612] Steven Shapin, "Discipline and Bounding: The History and Sociology of Science as Seen through the Externalism-internalism Debate," *History of Science* 30 (1992): 333-369. See also: Arie Rip, "Onderzoek van de Ontwikkeling van Wetenschapsgebieden," *Kennis en Methode* 1 (1977): 311-333.
[613] For further reading regarding scientific cultures: Benoit Godin and Yves Gingras, "What Is Scientific and Technological Culture and How Is it Measured? A Multidimensional Model," *Public Understanding of Science* 9 (2000): 43-58; Barry Barnes and David Edge, ed., *Science in Context: Readings in the Sociology of Science* (Milton Keynes: The Open University Press, 1982). For a concrete example: David Baneke, "Synthetic Technocracy: Dutch Scientific Intellectuals in Science, Society and Culture, 1880-1950," *British Journal for History of Science* 44 (2011): 89-113.
[614] Shapin, "Discipline and Bounding," 333.
[615] Johan A. Linthorst, "An Overview: Origins and Development of Green Chemistry," *Foundations of Chemistry* 12 (2010): 55-68. In part, this chapter builds further on this publication.

types of chemical research that became associated with the term green chemistry, and how they related to those concepts.

Below I will also discuss green chemistry as a scientific specialty, in order to pay attention to the social aspects of the field and assess whether green chemists form an intellectually closely connected community. As discussed in Chapter 1, scientific specialties are defined by Nicholas Mullins as 'institutionalised cluster[s]' with 'formal organisation'.[616] Related to that institutional organization, a scientific specialty is also characterized by the sharing of a paradigm between scientists, which connects them intellectually.

The philosopher Ernan McMullin questioned the radical nature of scientific revolutions.[617] He discerns three categories of "revolutions": deep, shallow and aesthetic. This perspective is employed too in this chapter in order to investigate, on the basis of internal scientific values, the significance of dominant (chemical) concepts that are associated with green chemistry.

The overall aim is to shed new light on the different views concerning the growth of green chemistry and in particular on the extent to which the intellectual origins shaped that growth. Is green chemistry, similar to for instance organic chemistry or physical chemistry, a "subfield" of the chemical science? Or is it a distinctive approach for greening the entire chemical enterprise? As I have shown in Chapter 1, STS-scholars came to diverging answers on these kind of questions. The leading questions to be answered in this chapter therefore are: which dominant chemical views and concepts on what green chemistry is and should be emerged and developed in the USA, the UK and the Netherlands? What were the goals that (influential) chemists tried to achieve when developing green chemistry? To what extent were these chemists intellectually connected? Ultimately, this chapter aims to understand the character of green chemistry as an activity and specialty, and the motives that helped to achieve scientific recognition for green chemistry.

I will first discuss the nature of green chemistry in the USA, where most of the new developments started.

---

[616] Nicholas C. Mullins, "The Development of a Scientific Specialty: the Phage Group and the Origins of Molecular Biology," *Minerva* 10 (1972): 56 and 74. See also: Thomas S. Kuhn, *The Structure of Scientific Revolutions* (London: The University of Chicago Press, 1996).
[617] Ernan McMullin, "Rationality and Paradigm Change in Science," in *Philosophy of Science: the Central Issues*, ed. Martin Curd and Jan A. Cover (New York, NY: W.W. Norton & Company, 1998), 119-138. See also: Thomas S. Kuhn, "Objectivity, Value Judgement, and Theory Choice," in *The Essential Tension: Selected Studies in Scientific Tradition and Change*, ed. Thomas S. Kuhn (London: The University of Chicago Press, 1977), 320-339.

5.1 USA: Conceptual Meanings of Green Chemistry

According to officers of the Environmental Protection Agency (EPA) there were different pollution prevention policies in the late 1980s and the early 1990s (see Chapter 2). In 1989, David Stephan and John Atcheson for instance argued that "waste minimization" fundamentally differed from "pollution prevention", because waste minimization solely concerned hazardous wastes, whereas pollution prevention concerned all wastes.[618] Therefore, 'EPA adopted the term pollution prevention to replace waste minimization'.[619] In 1992, Joseph Breen and Michael Dellarco defined pollution prevention as 'a combination of industrial source reduction and toxic chemical use substitution'.[620] The conceptual shift from "waste reduction" and "waste minimization" activities to pollution prevention policies had a strong impact on the emergence of green chemistry in the USA. In the next two sections I will likewise focus on the role played by the labels "benign by design chemistry" and "green chemistry". What did they mean, and what types of chemical activities were implied by these terms?

5.1.1 The Emergence of Benign by Design Chemistry

The term "benign by design chemistry" was first used by some EPA officers who tried to interest chemists for a new funding program launched by the EPA.[621] Starting in the early 1990s, some chemists benefitted from this program and lectured at symposia of the ACS Division of Environmental Chemistry (DEC). The publishing of *Benign by Design: Alternative Synthetic Design for Pollution Prevention* in 1994 was based on a symposium of DEC at the 206th National Meeting of the American Chemical Society (ACS) in August 1993.[622] This was the first book length publication on benign by design chemistry. All the authors were from the USA and three out of fourteen chapters were written by EPA representatives. The research by academic chemists reported in the other chapters was

---

[618] David G. Stephan and John Atcheson, "The EPA's Approach to Pollution Prevention," *Chemical Engineering Progress* 85 (1989): 53-58.
[619] Ibid., 54.
[620] Joseph J. Breen and Michael J. Dellarco, "Pollution Prevention: The New Environmental Ethic," in *Pollution Prevention in Industrial Processes: the Role of Process Analytical Chemistry*, ed. Joseph J. Breen and Michael J. Dellarco (Washington, DC: ACS, 1992), 6.
[621] See Chapter 2. Sometimes the term "benign chemistry" was used too within EPA, but far less than "benign by design chemistry". The meaning was considered equivalent by EPA officers though.
[622] Paul T. Anastas and Carol A. Farris, ed., *Benign by Design: Alternative Synthetic Design for Pollution Prevention* (Washington, DC: ACS, 1994).

financially supported by EPA's grant program, sometimes with additional funding provided by the National Science Foundation (NSF) or chemical companies. These scientists were the first to participate in this new EPA program. Three chapters were written by industrial chemists. According to the editors Paul Anastas and Carol Farris (both EPA policy officers) chapters in this book 'discuss the conceptual basis for benign chemistry and a wide range of research embracing this approach'.[623] They argued that the concept of benign by design chemistry was new and application oriented. Anastas and Farris attempted to persuade chemists in the laboratory and decision makers to embrace this new concept, which they defined – as shown in chapter 2 – as follows:[624]

> Benign by design chemistry is synthetic chemistry designed to use and generate fewer hazardous substances. The ultimate goal of benign chemistry research is to develop and institute alternative syntheses for important industrial chemicals in order to prevent environmental pollution.

This definition is very much in line with the "operational realism" character of chemistry in general.[625] A more comprehensive conceptual discussion on benign by design chemistry was presented in the chapter written by Anastas. He highlighted the role of synthetic chemists, which was 'crucial toward meeting the goals of both environmental protection and economic growth'.[626] Also the Pollution Prevention Act (PPA) tried to promote a close connection between practices of chemists in the laboratory and these environmental and economic goals.[627] Anastas observed that in developing new synthetic methodologies chemists mostly focus on yield. He admitted that this would remain of economic importance in chemical manufacture. But chemists were generally not concerned enough with environmental problems. This attitude, he argued, undermined the recognition by the public of the chemical enterprise:[628]

---

[623] Paul T. Anastas and Carol A. Farris, preface to *Benign by Design*, Anastas and Farris, IX.
[624] Ibid.
[625] See Chapter 1 and: Bernadette Bensaude-Vincent and Jonathan Simon, *Chemistry: The Impure Science* (London: Imperial College Press, 2008), 202-207.
[626] Paul T. Anastas, "Benign by Design Chemistry," in *Benign by Design*, Anastas and Farris, 2.
[627] "Pollution Prevention Act of 1990," in *Compilation of Selected Acts within the Jurisdiction of the Committee on Commerce: Environmental Law*, Committee on Commerce U.S. House of Representatives (Washington, DC: US Government Printing Office, 1995), 617-618; Anastas, "Benign by Design Chemistry," 5-10.
[628] Anastas, "Benign by Design Chemistry," 7.

The same industry, the chemical industry, that was once hailed as the provider of modern convenience and innovation, now is associated with fouling the planet.

Therefore new basic synthetic methodologies had to be developed that would address all these factors. This journey of 'innovation' was a task for synthetic chemists:[629]

Benign by Design chemistry defines synthetic elegance on the basis of three factors: efficiency of synthetic methodology; economically viable; environmentally benign.

So, Anastas was convinced that economic, environmental and chemical aspects could be combined and he therefore listed some general considerations for chemists, such as the type of reagents, solvents and feedstocks, in order to develop new synthetic methodologies to realize this. But, the reaction mechanisms, chemical strategies and concepts, characteristic of a synthetic methodology, were not really specified by him.[630] With one exception, namely the concept of "atom economy". Anastas considered this concept important for the practices and strategies of synthetic chemists, as it determined 'synthetic efficiency'.[631]

The concept "atom economy" had been coined by the synthetic organic chemist Barry Trost of Stanford University. In 1991 Trost reviewed in *Science* dozens of synthetic methodologies, which had (mostly) been developed in his own laboratory.[632] The efficacy of these methodologies was assessed in relation to selectivity, e.g., enantioselectivity, and to the number of steps of a synthetic pathway, the so-called "overall length". The latter should, of course, ideally be as small as possible for reasons of simplicity. Trost also mentioned scarcity of raw materials as a reason to be selective, and catalysis as a means to improve selectivity. In addition to these rather common sense considerations of synthetic chemists, he stated that another 'feature of [synthetic] efficiency is frequently overlooked – how much of the reactants end up in the product, a feature we might refer to as atom economy'.[633] This idea of atom economy is schematically explained below (Figure 5.1).

---

[629] Ibid, 7 and 10. Anastas mentioned in his chapter both industrial and academic chemists, including contributors to *Benign by Design*.
[630] For the general characteristics of synthetic methodologies, see: K.C. Nicolau and E.J. Sorensen, *Classics in Total Synthesis: Targets, Strategies, Methods* (New York, NY: VCH, 1996), 1-19.
[631] Anastas, "Benign by Design Chemistry," 10.
[632] Barry M. Trost, "The Atom Economy – A Search for Synthetic Efficiency," *Science* 254 (1991): 1471-1477. The majority of these synthetic methodologies was based on catalytic support of transition metals.
[633] Ibid., 1471.

In 1928 Otto Diels and Kurt Alder had developed a [2σ + 2σ + 2π] cycloaddition that was later called the Diels-Alder reaction, which has become a widely used synthetic methodology also employed by Trost.[634] Scheme 1 outlines this reaction between cyclopenta-1,3-diene (the diene) and methylvinylketone (the dienophile) in water. *All* the atoms of the diene and the dienophile are transformed into one single reaction product. This is the highest possible degree of atom economy as no by-products are formed. Scheme 2 contrasts with scheme 1. Following an electrophilic aromatic substitution mechanism, bromobenzene (desired product) is formed when benzene, bromine and the Lewis acid catalyst iron(III)bromide are mixed, but also hydrogen bromide (by-product) is formed.

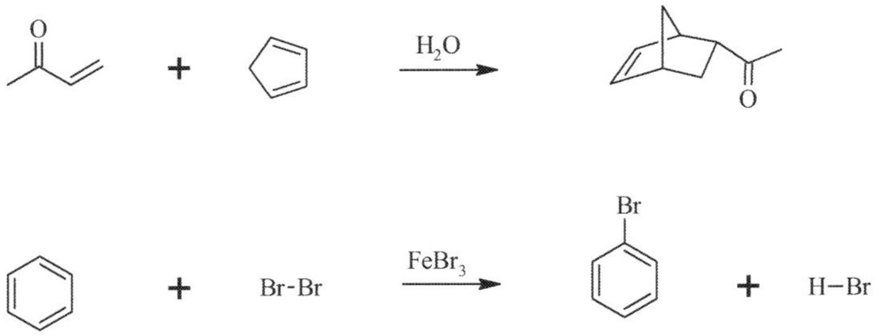

(1)

(2)

Figure 5.1: Reaction between cyclopenta-1,3-diene and methylvinylketone (**1**) and the synthesis of bromobenzene (**2**).[635]

Considering the atom economy concept of Trost, scheme 1 is favored over scheme 2, because:[636]

> The ideal reaction would incorporate all of the atoms of the reactants. Major benefits that derive from such processes include more effective use of limited raw materials and decreased emissions and waste disposal.

---

[634] Otto Diels and Kurt Alder, "Synthesen in der Hydroaromatischen Reihe," *Justig Liebigs Annalen der Chemie* 460 (1928): 98-122.
[635] Scheme 1: Wilfried Blokzijl, Michael J. Blandamer and Jan B.F.N. Engberts, "Diels-Alder Reactions in Aqueous Solutions. Enforced Hydrophobic Interactions Between Diene and Dienophile," *Journal of the American Chemical Society* 113 (1991): 4241-4246. The product obtained constitutes nearly 100 % endo selectivity. Scheme 2: Peter Vollhardt and Neil Schore, *Organic Chemistry: Structure and Function* (New York. NY: W.H. Freeman and Company, 2003), 670-671.
[636] Trost, "The Atom Economy," 1471.

Trost was thus aiming for synthetic methodologies that produced less pollution, in order to anticipate 'environmental concerns' of society.[637]

Atom economy was taken on board by EPA officers in order to persuade synthetic chemists to design new synthetic methodologies, in particular those to be employed in chemical industry, that would be in line with pollution prevention purposes. That's why Anastas did put forward atom economy as an evaluation criterion in his benign by design chemistry:[638]

> While full incorporation of atoms will never be achieved in all synthetic methodologies, it should be used as a goal and as one criterion by which to judge the benign nature of a reaction or reaction scheme.

What did all this imply for the practices in the laboratory? The chemists who published their research in *Benign by Design* tried to combine economic considerations with the growing public environmental concerns. For example, Karen Draths and John Frost expected synthetic routes *via* benzene 'likely to be in conflict with increasingly stringent environmental regulations' resulting in higher waste processing costs.[639] Therefore, they advocated the replacement of benzene, which originates from petroleum feedstocks, by D-glucose from renewable feedstocks in their synthesis of adipic acid, based in biocatalysis. Dennis Riley et al. discussed the replacement of phosgene in their organic synthesis of isocyanates and urethanes, because phosgene was on the list of 'extremely toxic substances' of the Industrial Chemistry Branch of EPA's Office of Pollution Prevention and Toxics (OPPT).[640] Clearly, the research agendas of these chemists, who built on previous research done in their laboratories, were influenced by growing environmental concerns and had different objects of study under consideration.[641]

---

[637] Ibid., 1477.
[638] Anastas, "Benign by Design Chemistry," 10.
[639] Karen M. Draths and John W. Frost, "Microbial Biocatalysis: Synthesis of Adipic Acid from D-Glucose," in *Benign by Design*, Anastas and Farris, 32.
[640] Dennis Riley, William D. McGhee and Thomas Waldman, "Generation of Urethanes and Isocyanates from Amines and Carbon Dioxide," in *Benign by Design*, Anastas and Farris, 122-132. See also: Carol A. Farris, Harold E. Podall and Paul T. Anastas, "Alternative Syntheses and Other Source Reduction Opportunities for Premanufacture Notification Substances at the U.S. Environmental Protection Agency," in *Benign by Design*, Anastas and Farris, 160; Michael K. Stern, "Nucleophilic Aromatic Substitution for Hydrogen: New Halide-free Routes for Production of Aromatic Amines," in *Benign by Design*, Anastas and Farris, 133-143.
[641] Authors often cited their own work from the 1980s, which is in advance of the benign by design chemistry concept.

But did the term and the idea of benign by design chemistry, as advocated by EPA officers, also have specific scientific, chemical implications? It is striking to note that none of the other authors in the volume did refer to the publications of Anastas and Trost, with the exception of George Kraus et al. They cited Trost and mentioned his concept of atom economy.[642]

None of the chemists that published in the volume, Anastas excepted, used the term benign by design chemistry in their scientific argumentation. Nevertheless, at the same time practicing chemists associated themselves with EPA's benign by design chemistry policy program, as illustrated by their involvement in the book *Benign by Design* and by the EPA research grants they received. Almost all chemists who published in the volume built on their own previous scientific work, without citing the publications of other contributors to *Benign by Design* though. Recalling Anastas and Farris call for action at the beginning of this section, which new 'conceptual basis' were these chemists in fact 'embracing' at this stage in their research?[643] It seems that a sort of ambivalence in recognition of benign by design chemistry existed that originates from a divide between contextual and internal scientific values. On the one hand, practicing (synthetic) chemists recognized the benign by design policy initiatives of the EPA due to public environmental concerns and the research grants they received (contextual).[644] On the other hand, these chemists generally were not citing Anastas or other contributors to *Benign by Design* and thus appeared to be weakly intellectually connected to each other via the benign by design concept. The chemical activities implied by this term appear to be rather diffuse.

5.1.2 A Changing Language: from Benign by Design Chemistry to Green Chemistry

Although none of the chemists that published in the volume, Anastas excepted, used the term benign by design chemistry in their scientific argumentation, there were two papers in which the new term "green chemistry" popped-up. Epling and Wang explicitly relate the use of the word "green" to green plants:[645]

---

[642] George A. Kraus, Masayuki Kirihara and Yusheng Wu, "A Photochemical Alternative to the Friedel-Crafts Reaction," in *Benign by Design*, Anastas and Farris, 76.
[643] Anastas and Farris, preface to *Benign by Design*, Anastas and Farris, IX.
[644] See Chapter 2.
[645] Gary A. Epling and Qingxi Wang, "Preparative Reactions Using Visible Light: High Yields from Pseudoelectrochemical Transformation," in *Benign by Design*, Anastas and Farris, 65. From a chemical and physical point of view, it is not right to call light a "reagent".

> Our approach (…) is to utilize visible light as the "reagent" which provides the thermodynamic driving force for a synthetic transformation. In this regard, the descriptor "green chemistry" is more than an accident, because the underlying mechanism that we rely upon for the key electron transfer step has a direct analogy to the photosynthetic mechanism utilized by green plants in the selective oxidation of water and reduction of carbon dioxide.

The authors of the second paper advocate the replacement of hazardous solvents by so-called supercritical CO_2. They emphasize that 'CO_2 is a *naturally-occurring* component of the atmosphere' and thus more environmentally friendly than 'any synthetic alternative', according to the authors.[646] The use of supercritical CO_2 represents in their view 'an important early step in the growth and development of "green chemistry"'.[647]

By appealing to the 'photosynthetic mechanism' in 'green plants' or considering CO_2 a 'naturally-occurring' substance, the authors that in the 1994 *Benign by Design* volume used the term green chemistry gave the term a meaning that related to the "imitation of nature".[648] Soon others would start using that term as well, but often with a different meaning.

In the mid-1990s, right after the publishing of *Benign by Design*, the US Presidential Green Chemistry Challenge Awards were announced.[649] A new terminology saw the light and, moreover, the term "green chemistry" became favored over "benign by design chemistry". But practicing chemists hardly used the new term at this stage, and mostly also without a clear scientific content. For instance, in 1995 Terrence Collins argued that green chemistry is 'real chemistry' that 'await clear definition', but comprises an 'anecdotal component', which means a power to look in 'a new and optimistic way' at chemistry and its future.[650] And in 1998 for David Kaplan et al. green chemistry was simply the chemistry that met 'environmental regulations'.[651] But in contrast with Collins and Kaplan, some EPA

[646] James M. Tanko, Joseph F. Blackert and Mitra Sadeghipour, "Supercritical Carbon Dioxide as a Medium for Conducting Free-radical Reactions," in *Benign by Design*, Anastas and Farris, 112.
[647] Ibid..
[648] See also Chapter 4. Starting in the early 1990s, in the Netherlands "green chemistry" became associated with "nature" too.
[649] See Chapter 2.
[650] Terrence J. Collins, "Introducing Green Chemistry in Teaching and Research," *Journal of Chemical Education* 72 (1995): 965-966.
[651] David L. Kaplan et al., "Enzymes in Polymer Science: An Introduction," in *Enzymes in Polymer Synthesis*, ed. Richard A. Gross, David L. Kaplan and Graham Swift (Washington, DC: ACS, 1998), 2-16. In 2003 Gross won as an academic chemist a Presidential Green Chemistry Challenge Award: "Green Chemistry Challenge Winners," EPA, accessed May 22, 2022, http://www2.epa.gov/green-chemistry/presidential-green-chemistry-challenge-winners#2003.

officers discussed the (chemical) views and goals of green chemistry in the 1990s in far more detail.

For instance, in 1996 Stephen DeVito and Roger Garrett (both EPA) published *Designing Safer Chemicals: Green Chemistry for Pollution Prevention*, which was based on a DEC symposium at the 208th ACS National Meeting in August 1994. They were employed by the OPPT office, from which the benign by design chemistry idea was promoted, and highlighted their 'Designing Safer Chemicals' concept, which was a 'sister concept under the Green Chemistry Program' of EPA.[652] The goal of this concept was the 'employment of structure-activity relationships (SAR) and molecular manipulation' in order to reduce negative toxicological effects of chemicals, according to Garrett in this volume.[653] He also argued that the idea of designing safer chemicals was not new, but the novelty was in the manner it was presented. It was supposed to serve as a conceptual framework.

Unlike other contributors to their book, DeVito and Garrett used the term green chemistry. For instance, DeVito formulated the goals of green chemistry:[654]

> Green chemistry strives to encourage the development of safer commercial substances and non-polluting commercial syntheses for purposes of reducing the risks posed by existing substances and traditional chemical syntheses.

He emphasized the importance of toxicological considerations in the (re)design of chemicals that should be employed by (industrial) synthetic chemists. He therefore transformed these considerations for simplicity purposes in a framework that comprised several toxicological questions for synthetic chemists, which should function as guidelines.[655] Garrett and DeVito stressed that this framework could be used in the laboratory. They also stressed the importance of governmental support to new research in this area in order to reduce the risks associated with chemical processes.[656] Despite of this toxicological wake-up call, this

---

[652] Roger L. Garrett and Stephen C. DeVito, preface to *Designing Safer Chemicals: Green Chemistry for Pollution Prevention*, ed. Roger L. Garrett and Stephen C. DeVito (Washington, DC: ACS, 1996), VII.
[653] Roger L. Garrett, "Pollution Prevention, Green Chemistry, and the Design of Safer Chemicals," in *Designing Safer Chemicals*, Garrett and DeVito, 5.
[654] Stephen C. DeVito, "General Principles for the Design of Safer Chemicals: Toxicological Considerations for Chemists," in *Designing Safer Chemicals*, Garrett and DeVito, 16.
[655] Ibid., 55.
[656] See for example: Garrett, "Pollution Prevention, Green Chemistry," 3.

scientific contribution to green chemistry was not cited very often.[657] So the impact on the field seems to have been limited.

Far more influential was Paul Anastas, who cooperated with dozens of coauthors. This illustrates his ability to create alliances in order to disseminate his conceptual ideas concerning green chemistry. For instance, in 1996 he published together with EPA officer Tracy Williamson *Green Chemistry: Designing Chemistry for the Environment*. This book was based, just like *Designing Safer Chemicals*, on a DEC symposium at the 208th ACS National Meeting.[658] Both authors continued EPA's benign by design chemistry efforts and cited chapters from *Benign by Design*, but a new term came to the fore in their new book: "green chemistry".

Anastas and Williamson provided several definitions of green chemistry. In line with "benign by design chemistry", they aimed to combine economic viability with a decreasing production of toxic wastes in the design of new chemical processes:[659]

> Green chemistry focuses on the design, manufacture, and use of chemicals and chemical processes that have little or no pollution potential or environmental risk and are both economically and technologically feasible.

According to Anastas and Williamson, the research published in their book stemmed from a new approach in the laboratory that was developed 'under the new paradigm of Green Chemistry'.[660] They argued that this approach 'received extensive attention and goes by many names including Green Chemistry, Environmentally Benign Chemistry, Clean Chemistry, Atom Economy and Benign by Design Chemistry'.[661] But the term green chemistry became now the most common one and was defined as:[662]

---

[657] According to the search engine Scopus (Accessed May 22, 2022), the chapter of DeVito has been cited 13 times and in the course of the late 1990s Garrett and DeVito were not very active in scientific publishing on green chemistry. By contrast, Anastas had many publications on green chemistry.
[658] Paul T. Anastas and Tracy C. Williamson, ed., *Green Chemistry: Designing Chemistry for the Environment* (Washington, DC: ACS, 1996). See also Chapter 2.
[659] Paul T. Anastas and Tracy C. Williamson, preface to *Green Chemistry: Designing*, Anastas and Williamson, XI.
[660] Anastas and Williamson, "Green Chemistry: An Overview," in *Green Chemistry: Designing*, Anastas and Williamson, 2.
[661] Ibid., 1.
[662] Ibid., 2. See for exactly one year later: Paul T. Anastas and Joseph J. Breen, "Design for the Environment and Green Chemistry: The Heart and Soul of Industrial Ecology," *Journal of Cleaner Production* 5 (1997): 97-102. See also Chapter 2.

> (…) the use of chemistry techniques and methodologies that reduce or eliminate the use or generation of feedstocks, products, by-products, solvents, reagents, etc., that are hazardous to human health or the environment.

They referred to chemical practices that were considered 'proven Green Chemistry methods and techniques', because these were 'more environmentally benign'.[663] That is an interesting phrasing, because instead of referring to 'chemical techniques and methodologies', as done in the quote, they now suggest that there are also 'green chemistry methods and techniques'. But what were these methods? According to Anastas and Williamson, there were two 'universal issues' in the field of green chemistry: atom economy and catalysis:[664]

> It is true that the principle of Atom Economy does not directly address the issue of hazard or toxicity. However, the incorporation of the goals of this paradigm is and needs to be one of the central tenants for any synthesis that is striving toward Green Chemistry. Another fundamental area of investigation that can effectively accomplish the goals of Green Chemistry is that of catalysis.

These 'universal' issues did not play a great role in the 1996 book, because similar to *Benign by Design*, none of the other contributors to *Green Chemistry: Designing* referred to work by Anastas, Trost or DeVito, but mainly cited their own previous research articles. The contributors employed different research approaches. For example, some chemists focused on the concept of 'Solid-State Polymerization of Amorphous Polymers'.[665] While others applied electron paramagnetic measurements, or discussed environmental concerns in inorganic chemistry education.[666] Apparently, the cognitive diversity in the types of research presented in *Green Chemistry: Designing* was huge.

Within a few years the situation would change considerably though. Owing to the efforts by the EPA officers mentioned the term green chemistry was used increasingly in scientific communication. Parallel to that the term green chemistry also got a more specific

---

663 Anastas and Williamson, "Green Chemistry: An Overview," 3.
664 Ibid., 14.
665 Kyosuke Komiya et al., "New Process for Producing Polycarbonate Without Phosgene and Methylene Chloride," in *Green Chemistry: Designing*, Anastas and Williamson, 20..
666 Angela Bassoli et al., "Oxidation of Phenolic Phenylpropenoids with Dioxygen Using Bis(salicylideneimino)ethylenecobalt(II)," in *Green Chemistry: Designing*, Anastas and Williamson, 92-104; James E. Huheey, "Incorporating Environmental Issues into the Inorganic Curriculum," in *Green Chemistry: Designing*, Anastas and Williamson, 232-238.

scientific content. For instance, when in 1996 the National Research Council recommended the intensification of research on “industrial ecology”,[667] Breen and Anastas responded:[668]

> DfE [Design for the Environment] and Green Chemistry are central to the development of the industrial ecology, *i.e.* systems approach to resolving this issue. It serves as a working example – understandable to the average citizen – of the connectedness of DfE-Green Chemistry and industrial ecology, as we move forward to make sustainable development a reality in the 21st century.

So Breen and Anastas connected in this statement green chemical practices in the laboratory to the wider issues of industrial ecology and sustainable development.

In 1998 Anastas and John Warner made a further step by publishing *Green Chemistry: Theory and Practice*. This frequently cited handbook started with the negative public image of chemistry, as there was ‘a large number of people who view chemicals and chemistry as something to be afraid of, curtailed, and avoided wherever possible’.[669] In order to avoid pollution and contribute to sustainability, green chemistry was the answer. This was now defined by Anastas and Warner as:[670]

> Green chemistry is the utilization of a set of principles that reduces or eliminates the use or generation of hazardous substances in the design, manufacture and application of chemical products.

What is new here is the highlighting of ‘a set of principles’ as a reflection of a green chemistry philosophy. Today those principles are best known as the “12 principles” introduced in the handbook by Anastas and Warner (Figure 5.2):[671]

> The listing of the “Twelve Principles of Green Chemistry” should be viewed as a reflection of the science that has been done within this nascent field in the recent past, as well as a direction

---

667 National Research Council, *Linking Science and Technology to Society’s Environmental Goals* (Washington, DC: National Academy Press, 1996).
668 Anastas and Breen, “Design for the Environment and Green Chemistry,” 102.
669 Paul T. Anastas and John C. Warner, *Green Chemistry: Theory and Practice* (New York, NY: Oxford University Press, 1998), 1. According to Scopus, this handbook has been cited approximately 7,730 times (Accessed May 22, 2022).
670 Ibid., 11.
671 Ibid., 29.

> that has been set by some of the pioneering scientists who have laid the groundwork for the future (…)

Many publications from the 1980s and the early 1990s by contributors to *Benign by Design* and *Green Chemistry: Designing* were cited in the handbook, and the "universal issues" mentioned above got a place too.[672] Concepts, ideas and scientific specialties, that have (partly) been mentioned above, all return as part of the 12 principles. Examples are: atom economy (principle 2), catalysis (principle 9), renewable feedstock's (principle 7) and toxicological, as well as public health considerations (principle 3).

1. It is better to prevent waste than to treat or clean up waste after it is formed.
2. Synthetic methods should be designed to maximize the incorporation of all materials used in the process into the final product.
3. Wherever practicable, synthetic methodologies should be designed to use and generate substances that possess little or no toxicity to human health and the environment.
4. Chemical products should be designed to preserve efficacy of function while reducing toxicity.
5. The use of auxiliary substances (e.g., solvents, separation agents, etc.) should be made unnecessary wherever possible and, innocuous when used.
6. Energy requirements should be recognized for their environmental and economic impacts and should be minimized. Synthetic methods should be conducted at ambient temperature and pressure.
7. A raw material of feedstock should be renewable rather than depleting wherever technically and economically practicable.
8. Unnecessary derivatization (blocking group, protection/deprotection, temporary modification of physical/chemical processes) should be avoided whenever possible.
9. Catalytic reagents (as selective as possible) are superior to stoichiometric reagents.
10. Chemical products should be designed so that at the end of their function they do not persist in the environment and break down into innocuous degradation products.
11. Analytical methodologies need to be further developed to allow for real-time, in-process monitoring and control prior to the formation of hazardous substances.
12. Substances and the form of a substance used in a chemical process should be chosen so as to minimize the potential for chemical accidents, including releases, explosions, and fires.

Figure 5.2: The 12 principles of green chemistry.[673]

672 Ibid., 125-129.
673 Ibid., 30.

But what is the scientific meaning and value of these principles? First of all, for centuries chemists have been using scientific laws, theories and synthetic methodologies in their scientific work. These differ in the number of users, but have profound similarities from a philosophy of science perspective, i.e., empirical adequacy in terms of accuracy, simplicity, scope and also consistency with other scientific theories and laws. A listing of principles (Figure 5.2), without specific mathematical formulae – such as the Lambert Beer law –, basic reaction mechanisms, or synthetic methodologies, is clearly a different matter (as far as chemical science is concerned).

This is illustrated by the fact that key proponents of green chemistry have been ambivalent in their expectations. Anastas and Warner characterized their new green chemistry approach as scientifically revolutionary.[674] And, in 2002, Anastas and Mary Kirchhoff (ACS Green Chemistry Institute) stated that 'green chemistry is intentionally designed, and that it is by definition impossible to do green chemistry by accident'.[675] In their view the 12 principles 'have been used as guidelines', and chemical research 'that strives to incorporate one or more of the 12 Principles of Green Chemistry' could be considered as 'excellent'.[676] The set of 12 principles obviously has a rather open character and is not compulsory for its (potential) users: chemists. So, there is a tension between the emphasis on the open character of green chemistry and the claim that green chemistry is no less than a scientific revolution.

The paragraphs above confirm the argument of Chapter 2 that the institutional origins of "green chemistry" in the USA are within the EPA and that it was disseminated by chemists such as Warner, Kirchhoff and Anastas. They attempted to persuade chemists to alter their practices and way of thinking. These efforts culminated in the 12 principles that should be employed in the laboratory. In 2002 Anastas and Kirchhoff concluded that green chemistry should not become a specific scientific discipline, but a new way of doing chemistry:[677]

> When the 12 Principles of Green Chemistry are simply incorporated as an integral part of everyday chemistry, there will no longer be a need for the focusing, highlighting, and moniker of green chemistry.

---

674 Ibid., 9 and 18. They called their green chemistry also a "philosophy".

675 Paul T. Anastas and Mary M. Kirchhoff, "Origins, Current Status, and Future Challenges of Green Chemistry," *Accounts of Chemical Research* 35 (2002): 686. In the 1998-2000 period Kirchhoff was in the EPA's Green Chemistry Program.

676 Ibid., 686 and 687. Anastas and Julie B. Zimmerman also employ that open character in their engineering framework to achieve sustainability. See: Paul T. Anastas and Julie B. Zimmerman, "Design through the 12 Principles of Green Engineering," *Environmental Science & Technology* 37 (2003): 95A.

677 Anastas and Kirchhoff, "Origins, Current Status," 691.

These proponents of green chemistry expressed thus a true scientific belief that the 12 principles could transform the chemical enterprise into a sustainable one. Such a transformation demands a strong internal scientific identity though, which makes the enterprise also vulnerable, when this is not realized.

To alter the way of thinking and doing of chemists with the help of a new paradigm, it must, in my view, include something fundamentally new and possess an internal scientific potential that has added values over the older theories and practices. According to Anastas and Williamson catalysis and atom economy could offer that basis. Catalysis already existed, but the concept of atom economy was new.[678]

According to Trost and Anastas, employing atom economy as a concept helps developing new chemical processes without producing toxic waste. The work of Trost has been frequently cited and received recognition by other chemists.[679] This also can be observed for *Green Chemistry: Theory* and the 12 principles of Anastas and Warner. This handbook has been cited frequently which might have been caused by a significant change in practices and ways of thinking by chemists. But it can also have been the result of the open character, or user-friendliness, of the 12 principles.

5.2 Green Chemistry in Europe

The democratization of environmental concerns in the UK and the Netherlands led to the institutionalization of new policies that impacted chemical science.[680] As a result of this, new terms such as green chemistry, clean technology and sustainable chemistry increasingly were used in these countries. The next sections will discuss the conceptual meaning of those terms in the UK and the Netherlands. I will investigate how deep the influence of the green chemistry concept of Anastas et al. was on the intellectual origins of green chemistry in these countries. According to Anastas and Warner, green chemistry is a new way of doing chemistry.[681] But, following Jody Roberts, 'ways of doing cannot be cleanly separated by ways of knowing, ways of thinking, or ways of constructing and analyzing experiments' in

---

[678] In fact, synthetic methodologies can be developed without acquaintance with the atom economy concept, though still realizing 100 % atom economy (e.g., the Diels-Alder reaction in 1928).
[679] According to Scopus, this most cited article of Trost has been cited approximately 3,950 times (Accessed May 22, 2022). It was hardly cited in the first years after publishing, which changed from 1996 onwards. Since then the citing behavior kept pace with the emergence of green chemistry.
[680] See Chapter 3 for the UK and Chapter 4 for the Netherlands.
[681] Anastas and Warner, *Green Chemistry: Theory*, 8-9 and 12-13.

the solving of scientific problems.[682] Therefore, I will also analyze to what degree a new paradigm from the USA has been shared internationally by green chemists.

5.2.1 The UK: from Clean Technology to Green Chemistry

As I have argued in Chapter 3 a key event in the UK was the launch of the journal *Green Chemistry* by the Royal Society of Chemistry (RSC) in 1999. James Clark, the founding editor of that journal, explained as the aim of the new journal:[683]

> The journal [*Green Chemistry*] is designed to appeal to a wide audience and will consist of a mixture of full papers, reviews, communications and news and comments from academe, industry, the government and the public.

It should become both a magazine and a 'high quality international' academic journal.[684] In the UK Clark was one of the most influential actors in the green chemistry field. How did he get that role and what was the impact of *Green Chemistry* on the intellectual development of green chemistry?

Clark, who was professor of chemistry at the University of York, referred in the late 1980s often to the negative environmental impact of industrial chemicals.[685] He searched for alternatives in order to minimize toxicity impacts as well as the amounts of waste. Based on these interests, he started using the term "clean technology" in the early 1990s, which he defined jointly with his colleague Duncan Macquarrie as follows:[686]

> Some of the major goals of clean technology in the chemical industry are to increase process selectivity, to maximise use of starting materials (aiming for 100 % atom efficiency), to

[682] Jody A. Roberts, "Creating Green Chemistry: Discursive Strategies of a Scientific Movement," (PhD diss., Virginia Polytechnic Institute and State University, 2005), 1-37. In VirginiaTechWorks; available from https://vtechworks.lib.vt.edu/handle/10919/27529, accessed May 22, 2022.
[683] James H. Clark, "The Greening of Chemistry," *Chemistry in Britain* 34 (1998): 45.
[684] James H. Clark, "Editorial," *Green Chemistry* 1 (1999): G1-G2.
[685] For example: James H. Clark et al., "Catalytic Oxidation of Organic Substrates Using Alumina Supported Chromium and Manganese," *Journal of the Chemical Society, Chemical Communications* 18 (1989): 1355-1356; James H. Clark et al., "Montmorillonite Supported Transition Metal Salts as Friedel-Crafts Alkylation Catalysts," *Journal of the Chemical Society, Chemical Communications* 18 (1989): 1353-1354.
[686] James H. Clark and Duncan J. Macquarrie, "Environmentally Friendly Catalytic Methods," *Chemical Society Reviews* 25 (1996): 303-310. See also Chapter 3.

replace stoichiometric reagents with catalysts and to facilitate easy separation of the final reaction mixture including the efficient recovery (and hopefully reuse) of the catalyst.

With 'aiming for 100 % atom efficiency', the societal objective of pollution prevention was emphasized and that required 'a level of innovation and new technology which the industry has not seen in many years'.[687]

According to Clark and Macquarrie this required new catalysts to be developed. They emphasized the disadvantages of traditional oxidation reactions that involved the use of toxic heavy metals in stoichiometric quantities. Consequently, they presented alternatives with a better environmental performance that were mostly developed in their own laboratory. These methods were often based on so-called "supported reagent type catalysts" that had a maximum dispersion and activity. The fixing of catalysts on a supporting medium was then a relatively new field of study, but it 'already proved [its] value in many organic reactions', according to Clark and Macquarrie.[688] They re-emphasized their ideas in 1998 in one of their most cited articles ever.[689] In this paper they also switched from "clean technology" to the term "green chemistry". They defined the latter as follows:[690]

> Some of the major goals of 'Green Chemistry' are to increase process selectivity, to maximise the use of starting materials (aiming for 100 % atom efficiency), to replace stoichiometric reagents with catalysts and to facilitate easy separation of the final reaction mixture including the efficient recovery of the catalyst.

This is almost the same definition as Clark and Macquarrie had given a few years earlier for "clean technology". The authors did not refer to any of the works by Anastas.[691] For Clark the terms clean technology and green chemistry obviously were interchangeable. The habit of mixing terms was continued by Clark in later work.

After launching his journal Clark did refer to Anastas and Warner though in his first editorial when he formulated the main objectives of *Green Chemistry*:[692]

---

[687] Ibid., 303.
[688] Ibid., 304 and 309. They emphasized the high "turnover numbers" of their catalysts.
[689] James H. Clark and Duncan J. Macquarrie, "Catalysis of Liquid Phase Organic Reactions Using Chemically Modified Mesoporous Inorganic Solids," *Chemical Communications* 8 (1998): 853-860. According to Scopus, this article has been cited approximately 420 times (Accessed May 22, 2022).
[690] Ibid., 853.
[691] James H. Clark, (phone) interview, December 12, 2010. Clark confirmed that his activities in the laboratory were independent from the terms chosen. See also Chapter 3.
[692] Clark, "Editorial," G1-G2.

Green chemistry is the utilisation of a set of principles that reduces or eliminates the use or generation of hazardous substances in the design, manufacture and application of chemical products.

So Clark used the same definition of green chemistry as Anastas and Warner had proposed in their handbook.[693] This suggests recognition of the concept of green chemistry as developed by these chemists from the USA.[694]

The journal *Green Chemistry* would regularly report on social matters. For example, Clark wrote about (international) concerns regarding the declining number of students who want to enroll in chemistry due to the negative public image of chemistry.[695] The news and comments sections of the journal reported on environmental issues, policy matters and industrial investments.[696] *Green Chemistry* had a huge impact on the increased use of the term green chemistry in academic journals since its launch in 1999, as demonstrated in Chapter 3.[697] In the first issue of this new journal Clark held a plea for 'new and successful "greener" chemistry' that would account for 'academic credibility and industrial relevance' in order to shape the 'green chemistry revolution'.[698]

Clark published in this first issue also one of his most frequently cited green chemistry articles ever, in which he discusses his views on the scientific identity of the concept of green chemistry.[699] The goal of preventing pollution was one of the key elements.[700] Clark connected this societal interest with his concept of 'ideal synthesis', which is 'a combination

---

[693] Anastas and Warner, *Green Chemistry: Theory*, 11. In addition, Anastas and Warner do not cite Clark in their handbook.
[694] Under editorial leadership of Clark, reflections on green chemistry activities, originating from EPA, were published. For example: James H. Clark, "Editorial," *Green Chemistry* 1 (1999): G29-G30; Paul T. Anastas, "Joe Breen – Heart and Soul of Green Chemistry," *Green Chemistry* 1 (1999): G87.
[695] James H. Clark, "Editorial," *Green Chemistry* 1 (1999): G117-G118.
[696] Ian D. Brindle, "Green Chemistry - A Canadian Perspective," *Green Chemistry* 1 (1999): G155-G157; Paul Anastas, Mary Kirchhoff and Tracy Williamson, "More 1999 Presidential Green Chemistry Awards," *Green Chemistry* 1 (1999): G174-G175; Anonymous, "Mercury Under Pressure," *Green Chemistry* 1 (1999): G4; Anonymous, "New Investments in the Czech Republic in Refinery," *Green Chemistry* 1 (1999): G3; Anonymous, "New Industrial Processes," *Green Chemistry* 1 (1999): G31-G32; Tim Lester, "Clean Technology Programme," *Green Chemistry* 1 (1999): G69-G71.
[697] Linthorst, "Origins and Development," 62-65.
[698] Clark, "Editorial," G1; James H. Clark, "Green Chemistry: Challenges and Opportunities," *Green Chemistry* 1 (1999): 1. In this article Clark also uses the term clean technology.
[699] Clark, "Green Chemistry: Challenges," 1-8. According to Scopus, this article has been cited approximately 730 times and is also among the most cited *Green Chemistry* articles (Accessed May 22, 2022).
[700] Ibid., 5-6. See for example: K. Wilson, A. Rénson and J.H. Clark, "Novel Heterogeneous Zinc Triflate Catalysts for the Rearrangement of α-Pinene Oxide," *Catalysis Letters* 61 (1999): 51-55.

of a number of environmental, health and safety, and economic targets' (Figure 5.3).[701] This idea of an "ideal synthesis" resonated with the overview he had presented in 1998 in his article written with Macquarrie. According to Clark 'two very broadly based generic areas of chemistry' had to be tackled in the future by chemists: 'acid catalyzed reactions' and catalyzed 'partial oxidations'.[702]

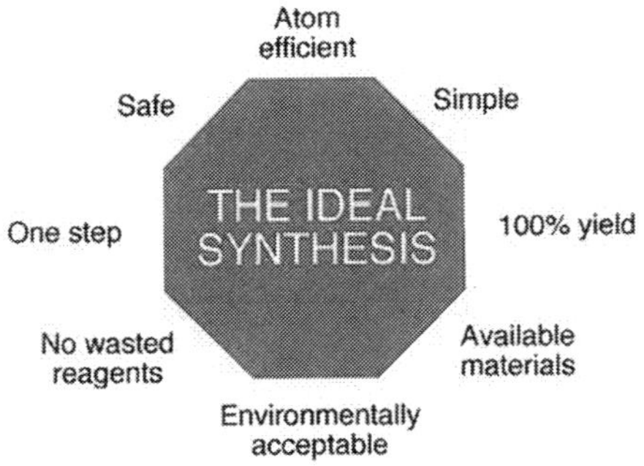

Figure 5.3: The ideal synthesis of Clark.

Comparing Clark's views on the ideal synthesis and on the two research areas of the future that he mentioned with the 12 principles of Anastas and Warner, we should notice that they are by no means a logical consequence from these 12 principles. Nor are they an alternative conceptualization of a green chemistry shaped by a natural law, a new scientific theory or paradigm, or a well-defined synthetic methodology. Instead, Clark stayed close within the realm of organic chemistry and catalysis. He concisely reviewed tens of synthetic methodologies by using the rhetoric of green chemistry and clean technology. Most of these methodologies had often been developed in his own laboratory and were the result of his earlier work.[703] Clark often compared traditional chemical processes with his own later ones. He advocated the new ones by highlighting their environmental qualities, such as the reduction of waste, or their higher atom efficiency. In this way his new (industrial) processes demonstrated the value of a green(er) chemistry, or clean technology. Clark thus built on

[701] Clark, "Green Chemistry: Challenges," 2.
[702] Ibid., 4.
[703] Ibid. In fact, Clark frequently cited his earlier works and did not mention the 12 principles of Anastas and Warner. See also: Clark and Macquarrie, "Catalysis of Liquid Phase Organic Reactions," 853-860; Clark and Macquarrie, "Environmentally Friendly," 303-310.

existing chemical concepts in catalysis, and his research agenda was also shaped by his interest in applied research.

With respect to the use of the term green chemistry, it appeared that Clark first labelled his own research program as clean technology. But later he chose the term "green chemistry".[704] Clark gave many examples of mostly catalytically based synthetic methodologies, which according to him were important for achieving environmental improvements in industry. These methodologies were labeled "green chemistry" or "clean technology", without making a clear distinction between those terms. Clark did not formulate a "new" chemical theory or paradigm to make chemistry greener along a well-defined scientific pathway. Instead, in his liberal view a great diversity of chemical research projects did all fall under the umbrella of green chemistry, which were published in his journal as we will see below.

5.2.2 The First Volume of *Green Chemistry*: a Scientometric Analysis

Green Chemistry has offered chemists the opportunity to publish research that reflected societal interests, but the articles in that journal also mirror a great scientific diversity. In order to investigate the character of that diversity, I have undertaken a scientometric analysis of all the 133 *Green Chemistry* articles in the first year of this journal, published by hundreds of authors from many different countries. My aim was to determine the degree of the so-called "bibliographic coupling" among the papers in that first volume. Bibliographic coupling means in this respect that those authors who cited in their *Green Chemistry* paper(s) the same document(s), share a reference pattern and are then bibliographically coupled. I have carried out my analysis by applying the procedure used by Loet Leydesdorff in several of his papers. By this methods he discovered and analyzed different networks, or clusters, of authors.[705] Those networks were visualized by using Pajek, a freeware Windows program, resulting in lines between authors who are bibliographically coupled (Figure 5.4).

---

[704] Clark, "Green Chemistry: Challenges," 1-8. The mixing of environmentally conscious terms is found in several *Green Chemistry* articles. See for example: T. Salmi et al., "Development of Clean Technology for the Production of Triols," *Green Chemistry* 1 (1999): 283-287; Brindle, "A Canadian Perspective," G155-G157.

[705] The *Green Chemistry* publications were, with the editorials included, obtained with Web of Science and used as input to run the file BibCoupl.EXE as prescribed by Leydesdorff. See: "Hands-on Class on Science & Technology Indicators," Loet Leydesdorff, accessed May 22, 2022, http://www.leydesdorff.net/indicators/lesson1.htm.

It shows that many authors, and clusters of authors, were rather isolated from other contributors to the journal. Apparently, authors from different articles in the first year of *Green Chemistry* built on a wide range of publications. As discussed in Chapter 1, this implies that these authors were concerned with rather different scientific puzzles. The clusters that emerged from my analysis consisted mainly of authors with a joint publication in one of the six issues of *Green Chemistry* in 1999. Some authors (who cited their own research) functioned as "nodes" by coupling with other authors (in another cluster). Examples are Rajender Varma and Keith Smith.[706] Most of the hundreds of *Green Chemistry* authors were weakly bibliographically coupled, and they were not cognitively united through the building on a small set of key publications on green chemistry. This is in line with the editorial style of Clark. Under his leadership many papers were published that concerned with quite different scientific puzzles, for instance the focus on ionic liquids as solvent for organic syntheses (Martyn Earle et al.), solvent-free organic synthesis (Varma), or supported reagent type catalysis (Clark). Clark also edited several special issues that originated from rather different knowledge domains, e.g., sonochemistry and renewable resources, but all under the umbrella of "green chemistry".[707]

---

[706] See for Varma for example: Rajender S. Varma, Kasi Pitchumani and Kannan P. Naicker, "Triphasic Catalyst Systems Based on Surfactant/Clay Composites. Facile Synthesis of Cyano, Thiocyano and Hydroxy Compounds Using a Triphasic Catalyst," *Green Chemistry* 1 (1999): 95-97; Rajender S. Varma, "Solvent-free Organic Syntheses Using Supported Reagents and Microwave Irradiation," *Green Chemistry* 1 (1999): 43-55. See for Smith for example: Keith Smith, Guy M. Pollaud and Ian Matthews, "Selective Production of 1-Arylalkenes. Synthesis of 1-Aryl-2-alkenes and 1-Arylalkanes *via* Friedel-Crafts Alkylation of Aromatic Substrates with Allylic Alcohols over Solid Acid Catalysts," *Green Chemistry* 1 (1999): 75-81; Keith Smith, Ping He and Ashley Taylor, "Selective *Para*-bromination of Phenyl Acetate under the Control of Zeolites, Bases, Acetic Anhydride or Metal Acetates in the Liquid Phase," *Green Chemistry* 1 (1999): 35-38.

[707] Martyn J. Earle, Paul B. McCormac and Kenneth R. Seddon, "Diels-Alder Reactions in Ionic Liquids: A Safe Recyclable Alternative to Lithium Perchlorate-diethyl Ether Mixtures," *Green Chemistry* 1 (1999): 23-25. Seddon received a Clean Technology Fellowship. The *Green Chemistry* issue of April 1999 concerned with renewable resources and the issue of June 1999 with sonochemistry.

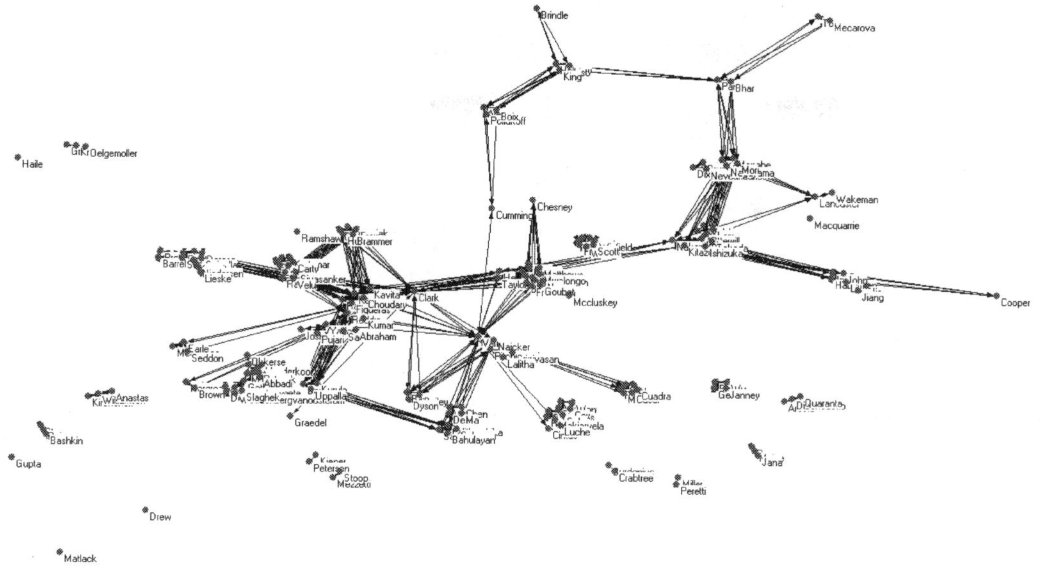

Figure 5.4: Bibliographic coupling among the *Green Chemistry* authors of 1999.

As argued above, in Clarks view a great diversity of chemical research projects did all fall under the umbrella of green chemistry. This diversity is clearly illustrated by the weak bibliographic coupling (cf., Figure 5.4) observed in *Green Chemistry*. Apparently, this major green chemistry journal was recognized by (green) chemists due to their similar societal interests, without being very restrictive as regards to the diverse chemical knowledge domains involved.

5.2.3 The Netherlands: Green and Sustainable Chemistry

As discussed in Chapter 4, in 1988 *Concerns for Tomorrow: A National Environmental Survey 1985-2010* [*Zorgen voor Morgen: Nationale Milieuverkenning 1985-2010*] was published by the Dutch National Institute for Public Health and the Environment [Rijksinstituut voor Volksgezondheid en Milieu] (RIVM).[708] Two of the main objectives of *Concerns for Tomorrow* were to achieve sustainable development and to provide an environmental outlook for the future. The report intended to inform and influence future

[708] F. Langeweg, ed., *Zorgen voor Morgen: Nationale Milieuverkenning 1985-2010* (Alphen aan de Rijn: RIVM, 1988).

Dutch science and technology policies, as well as policy in general. According to Barend van der Meulen, these policies were shaped on the basis of the highly decentralized Dutch policy making style of foresight, with stakeholders from industry, governments, science and society.[709] He discusses four large science policy programs of the 1990s, related to environmental concerns. One of these programs was Sustainable Technological Development [Duurzame Technologische Ontwikkeling (DTO)] (STD) that was divided into different research themes. Here I focus on the research program Sustainable Technological Development Chemistry [Duurzame Technologische Ontwikkeling Chemie (DTO-Chemie)] (STD-Chemistry) in order to discuss the discourse of the participants in that program concerning the use of the terms green and sustainable chemistry. I will compare the Dutch debate with the discourse on *Green Chemistry* in the UK, and the USA.

STD-Chemistry started in 1994. After some discussion concerning the relevance of chemistry within STD, eleven elite-experts of the Dutch chemical sciences were requested to present their personal vision on the role of the chemical sciences in the year 2040.[710] This took place in 1995 at a colloquium for a wide audience. STD-Chemistry built on the definition of sustainable development as formulated by the World Commission on Environment and Development, and strived for a 10 to 20 times more efficient use of energy and resources. This was pointed out in *Sustainability and Chemistry: A Collected Volume of Essays* [*Duurzaamheid en Chemie: Een Bundel Essays*] (1996), in which authors built on different chemical perspectives ranging from the chemistry of dyeing textile to agricultural production.[711]

Dominant perspectives concerned the banning of fossil feedstocks for energy consumption, the efficient use of (preferably) renewable feedstocks, and pollution prevention in the production of chemicals. For instance, Cees Okkerse and Herman van Bekkum strongly recommended an increased use of photo-voltaic energy and the use of biomass.[712] Despite the

---

[709] Barend van der Meulen, "The Impact of Foresight on Environmental Science and Technology Policy in the Netherlands," *Futures* 31 (1999): 7-23. See also: Barend van der Meulen and Arie Rip, "Mediation in the Dutch Science System," *Research Policy* 27 (1998): 757-769.
[710] See Chapter 4 for further reading on STD.
[711] DTO-Chemie, *Duurzaamheid en Chemie: Een Bundel Essays* (Delft: DTO, 1996).
[712] C. Okkerse and H. van Bekkum, "Vernieuwbare Grondstoffen voor de Chemische Industrie," in DTO-Chemie, *Duurzaamheid en Chemie*, 11-43. See also: Ludo van Klooster, "De Toekomst Ziet er 'Groen' en Zonnig uit," Delta: Journalistic Platform Delft University of Technology, last modified February 15, 1996, https://www.delta.tudelft.nl/article/de-toekomst-ziet-er-groen-en-zonnig-uit. Van Bekkum chaired the Royal Netherlands Chemical Society [Koninklijke Nederlandse Chemische Vereniging] (KNCV) in 1987-1988; "Bekkum, H. van", KNCV Chemie Historische Groep, accessed May 22, 2022, https://chg.kncv.nl/geschiedenis/biografieen/b/bekkum,-h.-van.

growing world population, they expressed, in line with the interest of major Dutch chemical industries and the Dutch Ministry of Economic Affairs, a firm belief in these approaches as they calculated for the year 2040 that fossil resources for energy consumption, and industrial production of organic compounds, could by then be completely phased out.[713] Okkerse and Van Bekkum listed ten widely used (industrial) organic compounds that already could be synthesized starting from biomass instead of fossil feedstocks.[714] They realized that their call for new knowledge and economic reform depended on the engagement of chemists with a program that focused on solar energy and renewable feedstocks, both in education and research. A new 'technological (r)evolution' was needed.[715] Jan Fortuin and Kees Boelhouwer discussed the reduction of waste in the production of high-value added chemicals.[716] Just like Okkerse and Van Bekkum, they recognized the importance of biomass, but also the concept "atom economy" as introduced by Trost.[717] They advocated the use of catalysts, e.g., organometallic and biocatalytic, in chemical syntheses for selectivity purposes. The use of nasty solvents also had to be reduced and replaced by environmentally friendly ones, such as water or supercritical fluids.[718] Apart from the engagement with biomass, the research approaches recommended by Fortuin and Boelhouwer differed from those of Okkerse and Van Bekkum, but there were profound similarities too.[719]

Firstly, the eleven elite-experts were all strongly committed to the belief that the chemical sciences could contribute to sustainability. Secondly, in line with the policy style of STD, a vision on the future was combined with examples of desirable research for demonstration purposes, in order to support the 'aggregation' of actors, ideas and interests

---

713 Okkerse and Van Bekkum, "Vernieuwbare Grondstoffen," 17. They referred to energy scenario calculations provided by the Shell company. For further reading on Dutch industrial and governmental interest in solar energy and biomass during the early 1990s, see: Ton van Helvoort and Ernst Homburg, *Een Eeuw Chemische Technologie in Nederland* (Delft: Stichting Hoogewerff-Fonds, 2021), 189-194.

714 See Van Bekkum also in a special issue of *Chemisch Magazine* on "green chemistry" ["groene chemie"]: A. Fuchs and H. van Bekkum, "Inuline, het Vergeten Zusje van Zetmeel," *Chemisch Magazine* 6/7 (1991): 320-321; H. van Bekkum and A. Fuchs, "De Verwachtingen Zijn Hooggespannen," *Chemisch Magazine* 6/7 (1991): 334-339.

715 Okkerse and Van Bekkum, "Vernieuwbare Grondstoffen," 42. All translations of quotations in this chapter are mine, unless otherwise stated.

716 J.M.H. Fortuin and C. Boelhouwer, "Nieuwe Reactiemedia voor de Fijnchemie in 2040," in DTO-Chemie, *Duurzaamheid en Chemie*, 45-105. For more background information concerning Fortuin and Boelhouwer, see: Ernst Homburg and Lodewijk Palm, ed., *De Geschiedenis van de Scheikunde in Nederland 3: De Ontwikkeling van de Chemie van 1945 tot het Begin van de Jaren Tachtig* (Delft: Delft University Press, 2004), 308.

717 Fortuin and Boelhouwer, "Nieuwe Reactiemedia," 74 and 103. They cited Trost.

718 Ibid., 47.

719 Principally, the authors of the different chapters did not cite each other works.

distributed over a wide range of institutions.[720] Thirdly and most notably, the contributors to *Sustainability and Chemistry* were not aiming at a theoretical conceptualization of the terms "green chemistry" or "sustainable chemistry".[721] Concrete examples formed the major content of the book. But in later stages of the STD-Chemistry program, reflection on the terminology did take place.

Several symposia, meetings and workshops were organized in later years. This led to publications for politicians and wider audiences in order to demonstrate that the chemical sciences could truly contribute to sustainability.[722] Moreover, in 1997 two comprehensive reports were published on the basis of STD-Chemistry that offered some conceptualization: *Sustainable Chemistry: Working Strategies for Sustainability in the Chemical Industry with a Reflection on Current Practices* [*Duurzame Chemie: Werkstrategieën voor Duurzaamheid in de Chemische Industrie met een Toetsing aan de Huidige Praktijk*] and *Sustainable Technological Development of Chemistry: Five Development Fields; Customized Research* [*Duurzame Technologische Ontwikkeling Chemie: Vijf Technologie Ontwikkelingsvelden; Onderzoek op Maat*].[723]

The first report was published by environmental NGOs united in a project entitled Sustainable Chemistry [Duurzame Chemie]. These NGOs cooperated with researchers of Dutch universities, e.g., Roger Sheldon (Delft University of Technology) as well as government representatives. There was also some communication with representatives of the STD-Chemistry program.[724] Based on analyses of widely used chemicals, the environmental NGOs argued that chemical industries had a long way to go in order to tackle Dutch environmental concerns.[725] The so-called "E factor" of Sheldon, an organic chemist with a

---

[720] Van der Meulen, "The Impact of Foresight," 21.
[721] Principally, the authors did not use these kind of terms in their reasoning.
[722] DTO-Chemie, *Reactie op de Toekomst: DTO-Chemie* (Amsterdam: Duurzame Technologische Ontwikkeling, 1997); DTO, *DTO Sleutel Chemie: Zon en Biomassa: Bronnen van de Toekomst* (Den Haag: Uitgeverij ten Hagen & Stam, 1997).
[723] Jan W. Biekart, Anton Nigten and Dick Stoppelenburg, ed., *Duurzame Chemie: Werkstrategieën voor Duurzaamheid in de Chemische Industrie met een Toetsing aan de Huidige Praktijk* (Utrecht: Stichting Natuur en Milieu, Vereniging Milieudefensie and Waddenvereniging, 1997); DTO-Chemie, *Duurzame Technologische Ontwikkeling Chemie: Vijf Technologie Ontwikkelingsvelden; Onderzoek op Maat* (Den Haag: Willems & Van den Wildenberg, 1997).
[724] Biekart, Nigten and Stoppelenburg, *Duurzame Chemie: Werkstrategieën voor Duurzaamheid*, 180. Leo Jansen (STD) and Sheldon also joined in 1991 the small KNCV taskforce "Commissie Chemisch Milieu Onderzoek" that was launched in order to identify environmentally engaged chemical research. See: KNCV, "KNCV Jaarverslag 1991," *Chemisch Weekblad* 88 (1992): 158. See also Chapter 4.
[725] Biekart, Nigten and Stoppelenburg, *Duurzame Chemie: Werkstrategieën voor Duurzaamheid*, 57-58.

focus on (bio)catalysis who became a productive green chemistry author in later years, was used in the NGO's report.[726]

In the early 1990s Sheldon had recognized the growing public environmental concerns and social pressures on chemical industries directed against the large production of waste in the production of fine-chemicals and pharmaceuticals.[727] He therefore introduced in 1992 the E factor. In contrast to the concept of atom economy of Trost (see section 5.1.1), which is a theoretical number that assumes 100 % yield in chemical synthesis, the E factor quantifies chemical processes in practice as 'the amount of waste produced per kilogramme of product'.[728] The E factor was defined by Sheldon in a simple formulae to calculate the environmental quotient as an indicator of environmental elegance for synthetic methodologies:

Environmental quotient = $E \times Q$

Here E is given by:

$$E = \frac{kg\ waste}{kg\ product}$$

and is Q the so-called unfriendliness quotient of the waste formed, i.e., high values of Q for environmentally unfriendly wastes. Sheldon argued:[729]

> For example, if innocuous salts such as NaCl are assigned a Q value of 1, then heavy metal salts, e.g., chromium, could be given a value of, say 100 or 1000, depending on toxicity. Such figures are obviously debatable and will vary from one company, or even one production site, to another.

---

[726] Ibid., 55. Sheldon was a colleague of Van Bekkum at Delft University of Technology. According to Scopus, Sheldon has been internationally among the most productive authors of green chemistry labeled publications (Accessed May 22, 2022).

[727] Roger A. Sheldon, "Organic Synthesis – Past, Present and Future," *Chemistry and Industry* 23 (1992): 903-906. According to Scopus, this article has been cited approximately 520 times (Accessed May 22, 2022). The citation frequency increased from the early 2000s onwards in particular, which ran parallel to the out of the blue growth of green chemistry. See Figure 3.2 in Chapter 3.

[728] Ibid., 904. See also: Roger A. Sheldon, "Selective Catalytic Synthesis of Fine Chemicals: Opportunities and Trends," *Journal of Molecular Catalysis A: Chemical* 107 (1996): 75-83; Roger A. Sheldon, "The E Factor: Fifteen Years On," *Green Chemistry* 9 (2007): 1273-1283; R.A. Sheldon, *E Factors, Green Chemistry and Catalysis: Records of the Travelling Chemist* (Delft: TU Delft, 2007), 4-9.

[729] Sheldon, "Selective Catalytic Synthesis of Fine Chemicals," 77. See also: Sheldon, "Organic Synthesis," 905.

Although Sheldon recognized the importance of 'atom utilisation' and toxicological considerations in chemical syntheses, he did not cite Trost, DeVito and Garrett, or Anastas. He often cited his own work though in his attempts to characterize and quantify the environmental impact of the type of waste formed.

The impact of waste influences for instance biodiversity and, accordingly, should be used in lifecycle analyses for pathways of elements and compounds in the environment. But according to the environmental NGOs, this was not sufficiently warranted within the STD-Chemistry program. It was argued that chemical industries needed guidelines in order to harmonize with the overarching objective of sustainable development.[730] The authors of the NGO report did not give a simple and concise definition of what the term sustainable chemistry means. Instead they presented five working strategies for chemical industries to accomplish the goal of sustainability:[731]

> 1. Ten to twenty times more efficient use of commodities and energy; 2. Reduction of using environmentally hazardous compounds; 3. Maximization of using renewable feedstocks; 4. Transformation to sources of renewable energy; 5. Harmonizing production with needs.

By highlighting the strategies three through five in particular, these NGOs stressed that much more should be done and attempted than was proposed by the representatives of the chemical industry who were involved in STD-Chemistry.

The second report was *Sustainable Technological Development of Chemistry*, which marked the end of the first period of the STD government policy program. In line with Dutch environmental policy in general, the idea of sustainable development was dominant in this report.[732] Next to that, research areas were identified that met future quality standards of living. At the same time the commitment of industry and the universities to new research agendas in aiming for sustainability was considered crucial.[733] The signaling function of the environmental NGOs for identifying environmental pollution as a social problem was

---

[730] Biekart, Nigten and Stoppelenburg, *Duurzame Chemie: Werkstrategieën voor Duurzaamheid*, 14.
[731] [1. Tien- tot twintigmaal efficiënter grondstof- en energiegebruik; 2. Afbouw van het gebruik van milieugevaarlijke stoffen; 3. Maximalisatie inzet van vernieuwbare grondstoffen; 4. Omschakeling naar bronnen van stromingsenergie; 5. Productie naar behoefte.] As in: Ibid., 53.
[732] World Commission on Environment and Development, *Our Common Future* (Oxford: Oxford University Press, 1987).
[733] DTO-Chemie, *Duurzame Technologische Ontwikkeling Chemie*, 4.

admitted. A long term vision for the chemical enterprise was preferred over the short term, and to that end "sustainable chemistry" was found a necessity.[734]

With respect to this new concept of "sustainable chemistry" it was argued in 1997:[735]

> The overlap between sustainable development and sustainable chemistry is much larger than it seems on first sight: chemistry is the field of changing matter and so is causally related to material sustainable development.

This illustrates that a scientific definition of the term sustainable chemistry was not aimed at, but, instead, it is a societally inspired claim that the objects studied by chemists were key to the issue sustainability.

The following five research areas were identified: (1) whole crop utilization [Geïntegreerde plantconversie] (GPC); (2) conversion of hydrocarbons [Koolwaterstoffenconversie] (C_1Chemistry); (3) photovoltaic solar energy [Fotovoltaïsche zonne-energie] (PZC); (4) fine chemicals process technology [Fijnchemische procestechnologie] (PFC); and (5) fiber-reinforced composite materials [Vezelversterkte composiet-materialen] (VVC).[736] Input from elite experts as mentioned in *Sustainability and Chemistry* returned in these five fields of study, but publications by Clark, Anastas and Warner were not mentioned. The three areas GPC, VVC and C_1Chemistry focused on biomass, PZC concerned the employing of solar energy, and PFC focused on pollution prevention. It was explicitly recognized that Dutch academic research concerning these five fields of study had already started, but its (policy) intensity had to be increased substantially.[737]

How did the Dutch developments relate to the international use of the term green chemistry and to the research done under that umbrella? Shortly after *Green Chemistry* started to appear, in 1999, also Dutch authors contributed to it. For example, in 1999 Marian C. Steverink-de Zoete et al. discussed an enzymatically catalyzed synthesis starting from a renewable feedstock like sucrose, and Hinderink et al. introduced three parameters in

---

[734] Ibid., 1 and 10. The authors referred to the report of the environmental NGOs.

[735] [De overlap tussen duurzame ontwikkeling en duurzame chemie is veel groter dan dat het op eerste gezicht lijkt: chemie is het vakgebied van de materie-verandering en heeft dus direct te maken met materiële duurzame ontwikkeling.] As in: Ibid., 9. Chapter 2 of this report concerns a concise theorizing of sustainable chemistry.

[736] Ibid. See Chapters 3 through to 7.

[737] Ibid., 1. See also: Van Helvoort and Homburg, *Een Eeuw Chemische Technologie in Nederland*, 190.

assessing sustainability through the lens of using renewable feedstocks, thermodynamic efficiency and the extent to which ecological cycles are closed down.[738] Although these authors were not among the key-actors in the STD-Chemistry program, these *Green Chemistry* articles show that Dutch chemists did associate themselves with green chemistry, even though they didn't use that term in their articles. They also did not cite Anastas, Warner of Clark. But they did focus on biomass, which corresponded with the Dutch focus on renewable resources as witnessed by the STD-Chemistry program.[739]

Okkerse and Van Bekkum, who were central to the STD-Chemistry initiative, were also among the first Dutch chemists who published in *Green Chemistry*. In 1999 Van Bekkum et al. wrote a short research paper on the renewable feedstock lactose.[740] The most important Dutch contribution to *Green Chemistry* in the same year was 'From Fossil to Green' though.[741] In comparison with other Dutch articles mentioned, this review article by Okkerse and Van Bekkum has been frequently cited. The title of their article also symbolizes important policy domains of STD-Chemistry, e.g., GPC and PZC. They argued:[742]

> In the year 2040 we will have to feed 9–10 billion people and to provide them with energy and materials. We want them to live according to the requirements of a developed society, and we do not want to pollute the earth or change the climate. The question addressed in this contribution is whether this scenario – a sustainable technological growth – is technologically possible on a world-wide basis.

In line with their Dutch contribution to *Sustainability and Chemistry*, Okkerse and Van Bekkum argued that the 'substitution of materials of today by "green" materials' was possible directly, or in a few reaction steps, from biomass.[743] In this respect and in line with C_1Chemistry policy, they expected methanol to develop into a 'key chemical' in order to

---

[738] Marian C. Steverink-de Zoete et al., "Enzymatic Synthesis and NMR Studies," *Green Chemistry* 1 (1999): 153-156; A.P. Hinderink, H.J. van der Kooi and J. de Swaan Arons, "On the Efficiency and Sustainability of the Process Industry," *Green Chemistry* 1 (1999): G176-G180.
[739] See Chapter 4 for the focus on renewable feedstocks. For a recent example of Dutch research on the replacement of fossil feedstocks by biomass, see: Christian A.M.R. van Slagmaat, et al., "Solvent-free Hydrogenation of Levulinic Acid to γ-Valerolactone Using a Shvo Catalyst Precursor: Optimization, Thermodynamic Insights, and Life Cycle Assesment," *Green Chemistry* 22 (2020): 2443-2458.
[740] A. Abbadi et al., "New Ca-sequestering Materials: Based on the Oxidation of the Hydrolysis Products of Lactose," *Green Chemistry* 1 (1999): 231-235.
[741] C. Okkerse and H. van Bekkum, "From Fossil to Green," *Green Chemistry* 1 (1999): 107-114. According to Scopus, this article has been cited 167 times and is the most cited Dutch 1999 article in this journal (Accessed May 22, 2022).
[742] Ibid., 107.
[743] Ibid., 107; Okkerse and Van Bekkum, "Vernieuwbare Grondstoffen," 11-43.

convert it into synthesis gas from which 'desired structures' can be build.[744] They stressed employing solar energy over the burning of fossil fuels as well.

Also Okkerse and Van Bekkum did not cite any publications by Clark, Anastas or Warner, nor did they use the term green chemistry in their review paper. Although it is interesting to note that Okkerse and Van Bekkum named renewable feedstocks 'green materials', in general they preferred the prefix "sustainable" over "green" in the 1990s. This is consistent with the findings in Chapter 4 concerning the STD-Chemistry program, in which green (chemistry) is associated with renewable feedstocks.[745]

The term sustainable chemistry as conceived in the Netherlands can be understood as a societal appeal in (the construction of) research agenda's to the efficient use of feedstocks, chemicals and chemical processes aiming at sustainability. That common aim, despite the differences in terminology, made publishing in *Green Chemistry* possible. Apparently, the Dutch papers matched the editorial publishing style and objectives of that journal. On the other hand, they also illustrate that the term "green chemistry" did not have a very fixed meaning, and could be exchanged by "sustainable chemistry" without a problem. This made the term green chemistry vulnerable, because there was 'considerable discussion with regard to what green chemistry embodies and whether, or not, the name itself is appropriate', wrote Sheldon in 2000.[746] In 2008, he expressed similar concerns in another editorial.[747] What is at stake here is the (vague) internal scientific identity of green chemistry, as I will argue in the next section.

5.3 The Scientific Foundations of Green Chemistry and the Controversy on the Terms "Green" and "Sustainable" Chemistry

The new terms "green chemistry", "clean technology" and "sustainable chemistry" increasingly came to the fore during the 1990s, which culminated in the establishment of the

[744] Okkerse and Van Bekkum, "From Fossil to Green," 107. A methanol molecule has one carbon atom, which explains the acronym C_1Chemistry.
[745] See also: DTO, *DTO Sleutel Chemie*, 7. And in the news magazine *C_2W* of the KNCV: Corry van Driel, "Groene Revolutie," *C_2W* 108 (2012): 5; Bastiënne Wentzel, "Chemicaliën uit Groenafval," *C_2W* 108 (2012): 20-23.
[746] Roger Sheldon, "Green Chemistry – One Year On," *Green Chemistry* 2 (2000): G1. See also: Roger Sheldon, "Aiming for that Goal of Zero Emission," review of *Green Chemistry: Theory and Practice*, by Paul T. Anastas and John C. Warner, *Nature* 399, no. 6731 (1999): 33; Sheldon, *E Factors, Green Chemistry and Catalysis*, 22.
[747] Roger A. Sheldon, "Green and Sustainable Chemistry: Challenges and Perspectives," *Green Chemistry* 10 (2008): 359-360.

journal *Green Chemistry*. In 1999 Walter Leitner argued that those terms were 'synonyms for the same approach'.[748] The next year Sheldon reported considerable discussion though concerning the scientific identity, as well as scientific recognition, of green chemistry. Although he admitted that it was hard to characterize the scientific identity of the field, he provided a 'reasonable working definition':[749]

> Green chemistry efficiently utilises (preferably renewable) raw materials, eliminates waste and avoids the use of toxic and/or hazardous reagents and solvents in the manufacture and application of chemical products.

His definition of green chemistry differed from the one given by Anastas and Warner in their handbook, nor did Sheldon refer to their 12 principles. In this section I will discuss how controversies help to get a better understanding of the empirical adequacy of green chemistry labeled research as far as its scientific identity is concerned.

Many countries increasingly adopted the policy aim of sustainable development in the course of the 1990s. As we have demonstrated above, new environmentally related terms were introduced within the chemical sciences during that period. This led to discussions and different preferences among chemists concerning the terms "green chemistry" and "sustainable chemistry" with respect to their use and conceptual meaning. For example, the Organisation for Economic Cooperation and Development (OECD) preferred the term sustainable chemistry. It organized for instance the OECD Workshop on Sustainable Chemistry in 1998, in which Warner and Anastas participated.[750] According to some people, especially from Germany, the prefix "green" held (left-wing) political connotations and in their view the terms "green chemistry" and "sustainable chemistry" could not be interchanged due to differences in meaning. That was also pointed out in 1999 by the prominent environmental chemist Otto Hutzinger who reported in an editorial on this controversy within

---

748 Walter Leitner, "Toward Benign Ends," *Science* 284 (1999): 1780-1781.

749 Sheldon, "Green Chemistry," G1.

750 Environment Directorate, *Proceedings of the OECD Workshop on Sustainable Chemistry: Venice, 15-17 October, 1998* (Paris: OECD, 1999). In 2004 the OECD organized the "Sustainable Chemistry – Integrated Management of Chemicals, Products and Processes" workshop in Germany. See also: Klaus G. Steinhauser, Steffi Richter and Jutta Penning, "Sustainable Chemistry in Dessau – A Workshop Report," *Green Chemistry* 6 (2004): G41-G43. In this workshop, 150 experts from 14 countries, representing industry, government, academia and NGOs, discussed sustainable chemistry. Anastas was also a participant.

the Division for Chemistry and the Environment of the Federation of European Chemical Societies. He argued:[751]

> The controversial debate has been about the use of the terms "Green Chemistry" or "Sustainable Chemistry" (...) In Germany, the term "green" in many elicits political-sociological feelings – for instance, fear of production shutdowns in chemical industry – and general opposition to genetic engineering projects and atomic energy by members of the Green Party. There has been strong opposition against the term "green chemistry" by IUPAC [International Union of Pure and Applied Chemistry], OECD, CEFIC [European Chemical Industry Council [Conseil Européen de l'Industrie Chimique]] and GDCh [the German Chemical Society [Gesellschaft Deutscher Chemiker]] headed by the German speaking members (...) The underlying meaning of the terms "Green Chemistry" and "Sustainable Chemistry" is different. Sustainable Chemistry is the maintenance and continuation of an ecologically sound development, whereas Green Chemistry focuses on the design, manufacture, and use of chemicals and chemical processes that have little or no pollution potential or environmental risk and are both economically and technologically feasible.

One year later Anastas responded in a meeting of the IUPAC Working Party on Synthetic Pathways and Processes on this controversy among chemists:[752]

> The terminology "green chemistry" or "sustainable chemistry" is the subject of debate. The expressions are intended to convey the same or very similar meanings, but each has its supporters and detractors, since "green" is vividly evocative but may assume an unintended political connotation, whereas "sustainable" can be paraphrased as "chemistry for a sustainable environment", and may be perceived as a less focused and less incisive description of the discipline. (...) The Working Party [on Synthetic Pathways and Processes] decided to adopt the term green chemistry for the purpose of this overview. This decision does not imply official IUPAC endorsement for the choice. In fact, the IUPAC Committee on Chemistry and Industry (COCI) favors, and will continue to use, sustainable chemistry to describe the discipline.

---

[751] Otto Hutzinger, "The Greening of Chemistry – Is it Sustainable?," *Environmental Science and Pollution Research* 6 (1999): 123. See for Hutzinger: Michael McLachlan et al., "Obituary Otto Hutzinger," *Environmental Sciences Europe* 25 (2013): 28; Chapter 4. See for the debate also: Pietro Tundo and Fabio Aricò, "Green Chemistry on the Rise: Thoughts on the Short History of the Field," *Chemistry International* 29 (2007): 4; Sheldon, "Green Chemistry," G1-G4.

[752] Pietro Tundo et al., "Synthetic Pathways and Processes in Green Chemistry. Introductory Overview," *Pure and Applied Chemistry* 72 (2000): 1207-1228.

Anastas' characterization of the term sustainable chemistry is in line with the definition presented by the OECD:[753]

> Sustainable chemistry is the design, manufacture and use of efficient, effective, safe and more environmentally benign chemical products and processes. Within the broad framework of sustainable development, government, academia and industry should strive to maximise resource efficiency through activities such as energy and non-renewable resource conservation, risk minimisation , pollution prevention, minimisation of waste at all stages of a product life-cycle, and the development of products that are durable and can be re-used and recycled.

But in Anastas' own view the terms green chemistry and sustainable chemistry obviously 'convey the same or very similar meanings', although the term sustainable chemistry 'may be perceived as a less focused and less incisive description of the [green chemistry] discipline'.[754] He and his immediate colleagues strategically positioned green chemistry as a unique field within the chemical sciences, closer to the laboratory as compared with sustainable chemistry. So the term sustainable chemistry was thus 'sometimes opposed and sometimes confused' with green chemistry.[755]

A similar controversy arose with respect to the 12 principles formulated by Anastas and Warner. In their view these principles were key for green chemistry, because they formed a conceptual framework that could be used as a scientific tool in the laboratory to achieve sustainability. This view was defended by Anastas and Warner, but was criticized in the early 2000s by Neil Winterton in *Green Chemistry*.[756] He argued that these principles were 'difficult to fault'.[757] Winterton urged for additional objectives and measures in order to assess the greenness of chemical processes, although Anastas and Warner had warned not to associate too high quantitative expectations with green chemistry:[758]

---

[753] Environment Directorate, *Need for Research and Development Programmes in Sustainable Chemistry* (Paris: OECD, 2002), 9.
[754] Tundo et al., "Synthetic Pathways and Processes in Green Chemistry," 1208.
[755] Gabriele Centi and Siglinda Perathoner, "Catalysis and Sustainable (Green) Chemistry," *Catalysis Today* 77 (2003): 287-297.
[756] Anastas and Breen, "Design for the Environment and Green Chemistry," 102.
[757] Neil Winterton, "Twelve More Green Chemistry Principles," *Green Chemistry* 3 (2001): G73-G75.
[758] Anastas and Warner, *Green Chemistry: Theory*, 15.

> (….) the goal of making a chemical product or process "environmentally benign" is a mere statement of the ethic of continuous improvement more than it is a metric by which to measure improvement.

Next to the 12 principles, they thus advocated a scientific ethic leading to a relatively open character for their green chemistry.

Despite that quite liberal view, during the *ad-hoc* growth of green chemistry in the late 1990s, several chemists at scientific symposia were searching for clarity: which types of chemical research count as green chemistry, and which ones not? After the 1999 conference Clean Processes and the Environment: The Catalytic Solution in Lyon (France) Sheldon, for instance, argued:[759]

> Subsequent lectures at the conference in Lyon generated considerable discussion with regard to what does and does not fall within the scope of green chemistry.

He held, in line with his E-factor, a plea for the use of semi-quantitative measures for assessing the greenness of chemical processes. Sheldon did not stand alone. In the literature an increasing number of green chemistry articles has been published (see Figure 5.5) that concerned the quantification of the greenness of chemical processes and products.

[759] Sheldon, "Green Chemistry," G2.

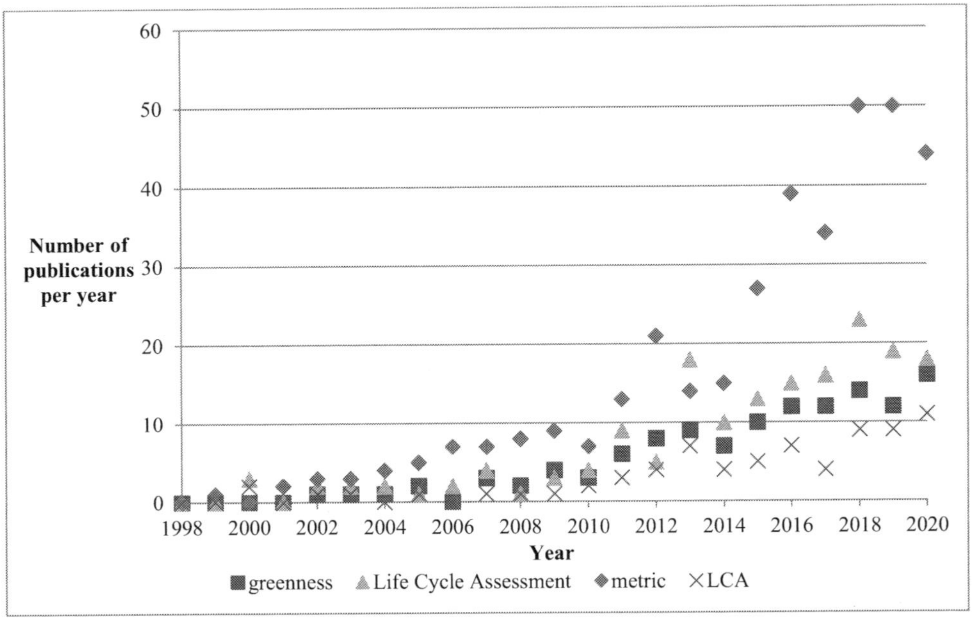

Figure 5.5: The growth of quantification approaches in "green chemistry" publications in which the terms metric, greenness, LCA or "life cycle assessment" are used.[760]

These publications are often closely related to the question of what counts as green chemistry, and what not. For instance, in 2005, Mark Peplow reported that ionic liquids do have 'an environmentally friendly reputation', but that these solvents might 'kill fish'.[761] While in the same year, Robin Rogers had won an EPA Presidential Green Chemistry Award for his academic green chemistry work on ionic liquids.[762] Contrasting research findings such as these ones, concerning the environmental friendliness of green chemistry labeled research, were summarized in 2005 as follows within the EPA:[763]

> Unfortunately, most new developments in green chemistry are not evaluated for their "greenness" via a systematic analysis of their environmental impact during their life cycle.

---

760 These figures were obtained by the author from ISI Web of Knowledge: Web of Science by searching on the "topic" green chemistry (Accessed August 8, 2021). The terms mentioned are used by the authors in "all fields" of their publications.

761 Mark Peplow, "Warning Shot for Green Chemistry," *Nature*, last modified November 3, 2005, http://www.nature.com/news/2005/051031/full/news051031-8.html.

762 EPA, "Presidential Green Chemistry Challenge Winners".

763 Yi Zhang, Bhavik R. Bakshi and Endalkachew Sahle-Demessie, "Evaluating the Greenness of Green Chemistry via Traditional and Thermodynamic Life Cycle Assessment," EPA, accessed May 22, 2022, https://cfpub.epa.gov/si/si_public_record_Report.cfm?dirEntryID=143141.

Consequently, it is likely that many techniques that are claimed to be green may not really have a smaller environmental impact over their life cycle.

So, since the start of green chemistry, the academic output of evaluating the greenness of green chemistry has been growing with considerable impact, as witnessed by the number of citations. This output mainly focused on employing the atom economy concept and the E factor.[764] While some authors also discussed lifecycle aspects and thermodynamics.[765] This trend within the chemical sciences to anticipate the assessment of environmental friendliness will probably continue to evolve.[766]

In a nutshell, there was international controversy on the meaning of the terms green and sustainable chemistry, which contributed to increasing efforts by chemists in order to assess the empirical adequacy of green chemistry labeled research (see Figure 5.5). This has proven to be a difficult task. Anastas and Warner therefore probably also relied on ethical values over demarcation metrics in order to give green chemistry an open character thereby aiming for alliances with chemists and policy makers. At the same time the analysis in this section illustrates that green chemistry is not the well-defined scientific field that some chemists and (STS) analysts assume it to be (see Chapter 1).

5.4 Discussion and Conclusions

During the early 1990s the term "green chemistry" was increasingly used in scientific publications. But how did this relate to the intellectual structure and scientific recognition of this green chemistry community? It appears that the intellectual origins of green chemistry are highly varied. The same applies to the scientific meaning of green chemistry, as a result of

[764] David J.C. Constable, Alan D. Curzons and Virginia L. Cunningham, "Metrics to 'Green' Chemistry – Which Are the Best?," *Green Chemistry* 4 (2002): 521-527. According to Scopus, this article has been cited approximately 570 times (Accessed May 22, 2022). Constable et al. presented in their frequently cited article the atom economy concept of Trost in a simple formula, which is based on the employing of the molar weights of reactants and the desired product in a chemical synthesis. See also: Marco Eissen et al., "Mass Efficiency as Metric for the Effectiveness of Catalysts," *Green Chemistry* 5 (2003): G25-G27.
[765] For example: Bernd Jastorff et al., "Progress in Evaluation of Risk Potential of Ionic Liquids – Basis for an Eco-design of Sustainable Products," *Green Chemistry* 7 (2005): 362-372.
[766] Stuart Cook, "Green Chemistry – Evolution or Revolution?," *Green Chemistry* 1 (1999): G138-G141; Tom Keijer, Vincent Bakker and J. Chris Slootweg, "Circular Chemistry to Enable a Circular Economy," *Nature Chemistry* 11 (2019): 190-195; Yvonne van der Meer, "How Sustainable Is It? That Is the Question," (Inaugural speech, Maastricht University, 2021), 1-20. In UM Research Publications; available from https://cris.maastrichtuniversity.nl/en/publications/how-sustainable-is-it-that-is-the-question, accessed May 22, 2022.

differences in rhetorical uses, absence of conceptualization, and of the way green chemistry was conceptualized by prominent green chemists.

As discussed in Chapter 1, the formation of a new scientific community starts traditionally with a few dominant scientists in the laboratory who are committed to 'paradigm development' and to solving similar research problems. This expands to a scientific specialty if the new paradigm is recognized.[767] A similar growth of a scientific community is not observed for green chemistry. For instance, the bibliographic coupling of articles published in *Green Chemistry* is weak (cf., Figure 5.4). This is caused by the fact that (green) chemists were devoted to different research problems and built on different intellectual resources. The example of Okkerse and Van Bekkum is a case in point. These two Dutch chemists were part and parcel of a policy program labeled "sustainable chemistry", which often concerns the efficient use of feedstocks.[768] Their views presented within this program were similar to those mentioned in their *Green Chemistry* article, in which they did not cite works of Clark, Anastas and Warner though. As pointed out for the Netherlands, the terminology of "sustainable chemistry" was largely rhetoric, and not characterized by a (fixed) paradigm. In the Netherlands the use of the term "green chemistry" was mainly reserved for the employment of renewable (bio)feedstocks.

In the UK, Clark has been one of the most productive and cited green chemists. Under his leadership the journal *Green Chemistry* published articles based on a wide range of intellectual perspectives and research problems, though in a spirit of sustainable development and improved environmental performance. Clark did not keep to a strict labeling, as he used the terms clean technology and green chemistry interchangeably in his own scientific work. Comparable to the Dutch case, Clark was not building on a (fixed) green chemistry paradigm. Nor did he present a new chemical theory under the umbrella of green chemistry.[769] He formulated "major goals of green chemistry", which culminated in his concept of "ideal synthesis", which was a combination of some important of social and ecological objectives

[767] Mullins, "The Development of a Scientific Specialty," 51-82.
[768] See for example: DTO-Chemie, *Duurzaamheid en Chemie*.
[769] See Chapter 1; Arie Rip and Jan-Peter Voss, "Umbrella Terms as Mediators in the Governance of Emerging Science and Technology," *Science, Technology & Innovation Studies* 9 (2013): 39-59; Bernadette Bensaude-Vincent, "The Politics of Buzzwords at the Interface of Technoscience, Market and Society: the Case of 'Public Engagement in Science'," *Public Understanding of Science* 23 (2014): 238-253.

such as the prevention of (toxic) waste. These goals were anecdotally highlighted by Clark with the listing of syntheses from his own laboratory (cf., Figure 5.3).[770]

The development of a new (green chemistry) paradigm that is incommensurable with an older one cannot be been observed for the Netherlands and UK. But an attempt was made in the USA.

Several policy initiatives were developed in the USA regarding pollution prevention. These activities were labeled "benign by design chemistry" in the early 1990s, but since the mid-1990s this changed to "green chemistry". Key actor was a man of the EPA office: Paul Anastas. He was one of the most productive green chemistry writers and has endlessly attempted to persuade chemists in the laboratory to employ his views on green chemistry in order to reduce the emissions of hazardous wastes. This philosophy, he argued, would improve environmental performance, through for instance pollution prevention, which would make economic sense too, leading to a sustainable development.

Advocates of green chemistry from the USA noted in 1999 that the growth of green chemistry 'has been driven by new knowledge'.[771] They did consciously choose and employ a new "green" theory in the laboratory. For instance, Anastas and Warner argued that 'right now the green chemistry revolution is beginning' and 'this is particularly appealing from a science-philosophy perspective'.[772] Anastas explicitly positioned green chemistry close to ways of doing and thinking of practicing chemists.

Together with Warner he published in 1998 a handbook with a conceptual framework, i.e., the 12 principles of green chemistry that could be used as guidelines (cf., Figure 5.2). They urged for an increased use of catalysis and atom economy in designing chemical syntheses. Catalysis was not new, but the atom economy concept was of a relatively recent origin.

In 2005 Jody Roberts stated that the 12 principles 'are also exclusionary, eliminating the participation of many through the strict technical focus of the principles', which could lead to 'solidification' of green chemistry.[773] Though the handbook of Anastas and Warner

---

770 Clark, "Green Chemistry: Challenges," 1-8. Kidwai and Mohan also employed this style of reasoning in order to underpin the fruitfulness of green chemistry labeled syntheses. See: M. Kidwai and R. Mohan, "Green Chemistry: An Innovative Technology," *Foundations of Chemistry* 7 (2005): 269 and 285.
771 Paul T. Anastas et al., "Promoting Green Chemistry Initiatives: Supported by a Rapidly Growing Infrastructure, the Field Promises Innovative Solutions to Pressing Environmental Problems," *Environmental Science & Technology* 33 (1999): 116A-119A.
772 Anastas and Warner, *Green Chemistry: Theory*, 9 and 18.
773 Roberts, "Creating Green Chemistry," 96.

has been frequently cited, because these principles are in my view not technical at all. In the first place, because they were not aimed at replacing traditional chemical views and practices, but at transforming them.[774] The 12 principles implied a certain openness to join green chemistry. In the second place, the 12 principles imply an user-friendliness, or an open character, due to the "should be sense" instead of a "must be sense". This open character was also enforced by the dominant spokespersons (Anastas and Warner) of green chemistry who relied on "scientific ethics" over metrics, as mentioned above. As a result, the scientific accuracy of the 12 principles, and therefore also the foundation of green chemistry, was contested, as witnessed by Winterton.

There is confusion with respect to terms such as green chemistry and sustainable chemistry, also as a result of political differences between countries.[775] This absolutely weakened the internal scientific profile of green chemistry, and this was therefore also subject of debate on the term and nature of green chemistry. This controversy did not really pose a serious barrier to the development of the most important journal, *Green Chemistry*, because this journal fostered from its start a very broad and liberal view of the new field. For instance, the early origins of green chemistry (USA) often concerned the reduction of toxic wastes, whereas sustainable chemistry (Nl) often appeared related to the efficient use of feedstocks. All those research done in these fields of study merged in *Green Chemistry* though.

Green chemistry was certainly not conceived as an abstract idea that was likely to bring 'order to phenomena' as a prerequisite to the choice of a new scientific theory or Kuhnian paradigm leading to a (deep) scientific revolution.[776] This is illustrated by the passive and more active rhetorical uses of the term green chemistry. The "passive uses" refer to the absence of any conceptual chemical meaning in rhetoric, or to scientific puzzling. So, to what Collins characterized as the 'anecdotal component'.[777] "Active uses", by contrast, refer to the conceptualization of green chemistry, for instance in the form of the 12 principles, or the "ideal synthesis" of Clark.

It appears to be difficult to design on paper a new (green) chemical process that is consistent with theories from the environmental sciences, such as "life cycle assessment" in order to aim for a better environmental performance. Epistemic beliefs and justifications concerning the value and scope of green chemistry thus differed, which led to an increased

---

[774] Anastas and Kirchhoff, "Origins, Current Status," 691.
[775] This is consistent with the findings of for example: Edward J. Woodhouse and Steve Breyman, "Green Chemistry as Social Movement?," *Science, Technology & Human Values* 30 (2005): 199-222.
[776] Kuhn, "Objectivity," 322; McMullin, "Rationality and Paradigm Change in Science," 119-138.
[777] Collins, "Introducing Green Chemistry," 965.

attention for attempts to assess the greenness (see Figure 5.5). This illustrates the scientific vulnerability of green chemistry.

Chapter 6: The Emergence of Green Chemistry: Discussion and Conclusions

The objective of this chapter is to answer the central research question "how did environmental concerns and the public image of chemistry influence the emergence and (scientific) development of green chemistry in the USA, the UK and the Netherlands?" To answer that question I have investigated which (political, governmental, environmental, etc.) factors and actors were dominant in these countries with respect to the birth and development of green chemistry in the course of the 1990s, and also in relation to the internal scientific shaping of the field.

Below, I will present a comparative analysis in order to identify similarities and differences among the three countries investigated, in particular with respect to the impact that environmental concerns and the public image of chemistry had on the emergence of environmental and green chemistry.

Section 6.1 discusses how these factors influenced policy making and how policies in their turn had an impact on the emergence of different types of green chemistry in the three countries. That section also discusses who were the dominant local advocates of green chemistry.

In the next section (6.2), the main ideas and guiding principles that were proposed are discussed in order to improve our understanding of the scholarly recognition of green chemistry and of the rhetorical uses of this term. This section also analyzes to what extent the emergence of green chemistry was influenced by new knowledge and how this differed between the three countries. After that comparative analysis I will focus in greater detail on the conclusions we can draw on the 'identity' of green chemistry, as well as on the drivers behind the growth of that new field.[778] As discussed in Chapter 1, scholars from various backgrounds came to different characterizations of green chemistry, ranging from substantial epistemic components to merely rhetoric's or even greenwashing. I will give a new characterization of the field and its scientific content and scope.

The concluding remarks of these sections are combined in section 6.3 in order to assess the meaning and relevance of green chemistry for science and society since its

---

[778] Laurens K. Hessels and Harro van Lente, "Practical Applications as a Source of Credibility: a Comparison of Three Fields of Dutch Academic Chemistry," *Minerva* 49 (2011): 215-240. See also Chapter 1.

emergence during the 1990s. The limitations of my research approach are discussed, finally, in section 6.4.

6.1 The Emergence of Green Chemistry: Different Origins in Countries with Different Political Cultures

The Chapters 2 through 4 describe the development of tensions between increasing public environmental concerns and the state of the economy. Parallel to that the membership of the environmental chemistry divisions of the three chemical societies that I investigated grew (see Figure 6.1).[779] The growth of this rather new scientific subfield, which peaked in the mid-1990s, illustrates the growing attention paid to the environment by chemical societies, and also by politics.[780] In terms of the relative share of the environmental chemistry division within the corresponding chemical society as a whole, the KNCV Division Environmental Chemistry [Sectie Milieuchemie] (hereafter MC) outnumbered the ACS Division of Environmental Chemistry (DEC) and the RSC Environmental Chemistry Group (ECG). This illustrates the supposed huge sociopolitical impact of Dutch environmental concerns on chemical science, as compared to the USA and the UK. This picture should be qualified though by the fact that the ACS had more technical divisions devoted to the environment than the KNCV and RSC, due to increased scientific specialization in the USA (Chapter 2). But obviously, changing contextual values impacted the structure of chemical societies and the identity of chemical science.

---

[779] The American Chemical Society (ACS), the Royal Society of Chemistry (RSC) and the Royal Netherlands Chemical Society [Koninklijke Nederlandse Chemische Vereniging] (KNCV).
[780] Specialization regarding environmentally engaged chemical research increased from the 1970s onwards, which led to the establishing of different subject groups and membership competition among these groups. Most likely, I conclude, specialization provides a minor argument for the decrease in environmental chemistry membership during the late 1990s.

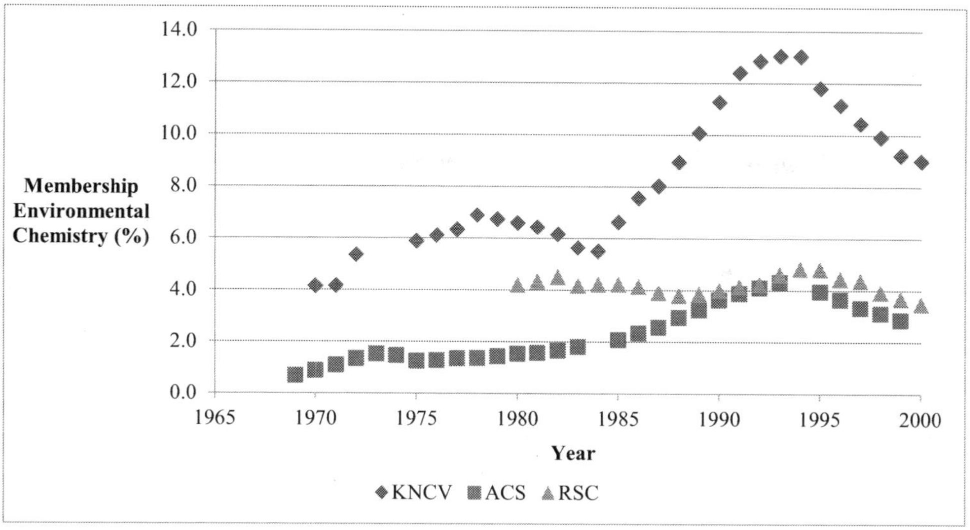

Figure 6.1: Membership of the environmental chemistry divisions within the ACS, RSC and KNCV, as percentage of the total membership of these societies.[781]

The increasing attention given to the environment during the 1990s went together with a growing negative public image of chemistry. But also demands for new chemical knowledge played a role. Applying the concept of the science-society contract proposed by Laurens Hessels and Harro van Lente, these factors changed the 'rationales' and 'conditions' of chemical science.[782] This coincided with the rise of green chemistry. It should be emphasized though, that the emergence and institutional origins of green (and sustainable) chemistries in the USA, the UK and the Netherlands developed along different historical lines and were based on different rationales, due to contrasting culturally conditioned styles of policy making. This in turn marks the value of comparative analysis in order to understand the impact of political culture on the emergence of green chemistry.

Chapter 2 discusses growing public environmental concerns in the USA since the 1970s, which resulted in a 'cascade of laws', command-and-control policies and increasing numbers of administrative and civil enforcement actions.[783] In this political culture, fueled by

[781] For the ACS: data are received from the ACS Office of Public Affairs. For the RSC: data are obtained from the RSC annual reports (1980-1994) and the RSC annual records (1995-2000). For the KNCV: data are received from the KNCV administration office.
[782] Hessels and Van Lente, "Practical Applications as a Source of Credibility," 217-218.
[783] William D. Ruckelshaus, "Environmental Protection: a Brief History of the Environmental Movement in America and the Implications Abroad," *Environmental Law* 15 (1985): 455-469;

pluralism and distrust among stakeholders, the US government relied on a legally driven style of policy making in order to establish public credibility and to tackle environmental problems. As a result, the (legal) polarization between government, chemical industries and society strongly peaked during the late 1980s. Characteristic of USA's democracy, a 'contentious' civic epistemology had thus developed.[784]

In the same period though a new environmental policy objective emerged within this civic epistemology in relation to the growing polarization: pollution prevention. In 1990 US Congress passed the Pollution Prevention Act (PPA). This environmental act changed the rationale for chemical science and spurred the emergence of green chemistry, which thus started in the USA with a clear focus on pollution prevention in order to reduce the production of toxic wastes.[785] First within the US Environmental Protection Agency (EPA) in particular. The PPA emphasized the economic and environmental drawbacks of command-and-control policies and published about industrial pollution reduction success stories in order to water down the polarization.[786] As a result, some dedicated public servants – and chemists by training – such as Paul Anastas and Joseph Breen of the EPA Office of Pollution Prevention and Toxics (OPPT), disseminated the PPA message extensively within DEC, a scientific network familiar to EPA officers. Breen also became the first executive director of the Green Chemistry Institute (GCI), which was created in 1997 as an autonomous non-profit institute with no physical location (Chapter 2). Anastas and Breen received the recognition of good-willing chemists to tackle environmental problems. Their influence was also great as a result of the fact that they could offer R&D funding to research on the design of chemical syntheses aiming at pollution prevention, with the help of the Design for the Environment (DfE) program of EPA.[787] This policy program received political recognition by the Bill Clinton/Al Gore Administration that announced in 1995 the Presidential Green Chemistry Challenge

---

Andrew J. Hoffman, *From Heresy to Dogma: an Institutional History of Corporate Environmentalism* (San Francisco, CA: The New Lexington Press, 1997), 89 and 113.

784 Sheila Jasanoff, *Designs on Nature: Science and Democracy in Europe and the United States* (Princeton, NJ: Princeton University Press, 2005), 259 and 262. See Chapter 1 for civic epistemology.

785 For example: Paul T. Anastas, "Benign by Design Chemistry," in *Benign by Design: Alternative Synthetic Design for Pollution Prevention*, ed. Paul T. Anastas and Carol A. Farris (Washington, DC: ACS, 1994), 2-22.

786 "Pollution Prevention Act of 1990," in *Compilation of Selected Acts within the Jurisdiction of the Committee on Commerce: Environmental Law*, Committee on Commerce U.S. House of Representatives (Washington, DC: US Government Printing Office, 1995), 615-623.

787 Anastas, "Benign by Design Chemistry," 2-22; Paul T. Anastas, "Design for the Environment: Chemical Design Project; Changing the Way Chemists Think about Chemicals," (AL: probably early 1993): 1-2 [Internal EPA communication].

Awards.[788] These awards helped to create 'publicity', and also reduced societal polarization with respect to chemistry, by showing good chemical practices to society.[789] This was in line with EPA's Green Lights policy program that also used the term green in order to enhance the 'environmental image'.[790]

The Presidential awards were embraced by the ACS, in cooperation with the EPA, in order to improve the public image of chemistry. The awards were used as a rhetorical tool at the cutting-edge of science and society as a means to improve the identity and legitimacy of chemistry and to receive more R&D funding from government.[791] In 2000 the recognition of the emerging field of green chemistry increased further when the GCI was absorbed by the ACS, with the aim to influence 'policymakers' and 'science policy'.[792] Obviously, the ACS considered the term "green chemistry" to be the right rhetorical instrument to secure the conditions for the discipline. This was an important reason for ACS to institutionalize green chemistry within its headquarter in Washington, DC.

Chapter 3 shows UK's public environmental concerns that received sociopolitical effects in the form of new legislations and institutional reforms by the UK government. In order to obtain public credibility, command-and-control policies were developed top-down by the state, but with a keen interest for UK's industrial competitiveness. The British policy making process was traditionally dominated by a small set of public servants and elite experts of high standing. This characterized UK's 'communitarian' civic epistemology.[793] But in practice, the government proved reluctant as 'very few prosecutions occurred' due to a 'partnership between the regulator and industry'.[794] The role played by environmental concerns within UK's political culture was relatively weak as compared to the USA and the

---

[788] Bill Clinton and Al Gore, *Reinventing Environmental Regulation* (Washington, DC: National Performance Review, 1995).

[789] Ibid., 42.

[790] EPA, *Green Lights: Third Annual Report* (Washington, DC: US Environmental Protection Agency, 1994), 4. See also: Hoffman, *From Heresy to Dogma*, 112 and 133.

[791] ACS, *American Chemical Society Annual Report 1996* (Washington, DC: ACS, 1997), 3. Federal government R&D expenditures to chemical sciences were on the decline, see: Gregory J. Hather et al., "The United States of America and Scientific Research," *Plos One* 5 (2010): 1-9.

[792] ACS, *Our 125th Anniversary: Charting our Course for the 21st Century: American Chemical Society Annual Report 2001* (Washington, DC: ACS, 2002), 3 and 12.

[793] Jasanoff, *Designs on Nature*, 259-261.

[794] James Connelly and Graham Smith, *Politics and the Environment: From Theory to Practice* (London: Routledge Taylor & Francis Group, 2003), 300.

Netherlands.[795] As shown in Chapter 3, this was especially important during the 1980s as a result of the economic policies of the Administration of Margaret Thatcher.[796]

The political attitude towards environmental problems in the UK changed in the late 1980s though. The European Economic Community put pressure on the UK government to embrace pollution control and to improve public transparency in the shaping of environmental policies.[797] As a result, the rationale for chemical science in relation to these issues changed. The result was an increasing focus on pollution prevention and clean technologies, as witnessed by *This Common Inheritance: Britain's Environmental Strategy* (1990).[798] Environmental objectives, and the poor public understanding of science, also came to the fore, especially in the science policy program Technology Foresight (TF) that started in the early 1990s under leadership of the UK's Chief Scientific Adviser.[799] The start of TF was a clear response to foresight activities undertaken in other countries, such as the Netherlands. Its aim was to secure UK's industrial competitiveness. UK's science policy had in comparison with other countries a more explicit 'market-driven approach'.[800] As shown in Chapter 3, this approach was only partially embraced by the RSC, because the society feared that it would undermine research on pure chemistry.[801]

Inspired by EPA's Presidential Green Chemistry Challenge Awards the RSC also decided to recognize the importance of green chemistry.[802] Armed with the rhetorical power of this term, RSC was able to demonstrate the crucial function of chemical knowledge in

---

[795] Ibid., 100; Derek Osborn, "Some Reflections on UK Environment Policy, 1970-1995," *Journal of Environmental Law* 9 (1997): 3-22. See also Figure 6.1.

[796] Mary O'Mahony, "Employment, Education and Human Capital," in *The Cambridge Economic History of Modern Britain*, ed. Roderick Floud and Paul Johnson (Cambridge: Cambridge University Press, 2004), 116.

[797] Secretaries of State for Environment, Trade and Industry, Health, Education and Science, Scotland, Transport, Energy and Northern Ireland, the Minister of Agriculture, Fisheries and Food and the Secretaries of State for Employment and Wales, *This Common Inheritance: Britain's Environmental Strategy* (London: HMSO Publications, 1990), 12.

[798] Ibid., 186.

[799] Office of Science and Technology, *Progress through Partnership: Report from the Steering Group of the Technology Foresight Programme 1995* (London: HMSO Publications, 1995), 80-82; Advisory Council on Science and Technology, *Technology Foresight: The Identification and Promotion of Emerging and Generic Technologies* (London: HMSO Publications, 1994); Luke Georghiou, "The UK Technology Foresight Programme," *Futures* 28 (1996): 359-377.

[800] Office of Science and Technology, *Progress through Partnership*, 22.

[801] Paul Stoneman, "Government Spending on Research and Development in the UK," *Fiscal Studies* 20 (1999): 223-259.

[802] John Brophy, "A Major New Initiative in Environmental Chemistry: Proposal for a Royal Society of Chemistry Green Chemistry Network," *Environmental Chemistry Group Newsletter* 8 (1998): 2-3. Clark attended in 1997 a green chemistry lecture of Anastas. See Chapter 3.

fulfilling the new knowledge demands of society as proposed within TF.[803] A crucial role within the TF initiatives and the emergence of green chemistry in the UK was played by the chemist James Clark, professor at York University. As a result, in 1998 the RSC Green Chemistry Network (GCN) was established, operating from York University. This helped in improving the public image and thus the conditions for the chemical sciences.[804] In comparison to the GCI and the Presidential Green Chemistry Challenge Awards in the USA, the institutional origins of GCN were thus not within an environmental agency, nor were its daily operations run within RSC's headquarter. In the case of the UK, a practicing chemist (Clark) had the lead. He had been committed to pollution prevention and was involved in the UK Clean Technology Programme already for several years.[805] He also became the first editor of the RSC journal *Green Chemistry*, which facilitated an explosive international growth of the field of green chemistry since its establishment in 1999. Clark followed a pragmatic editorial style. He published "green chemistry" labeled articles with the aim to tackle environmental problems. Clark was fully aware of the bad public image of chemical science and of the rhetorical advantages of the term green chemistry:[806]

> The title *Green Chemistry* was itself the subject of considerable discussion and debate. We considered many alternatives, but none carried the same combination of widespread use and appreciation, as well as simplicity and impact. (…) On the eve of the 21st century, however, the public are more aware of the hazardous substances that many chemical processes use and generate than the benefits of the products. Chemistry and the chemical industry have tarnished images. With your help we can use this journal to convey the means and motivation for chemists to make a difference.

Chapter 4 discusses the public environmental concerns in the Netherlands, and its consequences for environmental and green chemistry. The Dutch government responded to these problems in the early 1970s with new legislations and institutional reforms. It belonged

---

803 Anonymous, "The RSC's Scientific Forward Look and the UK Foresight Programme – Recent Developments," *Environmental Chemistry Group Newsletter* 9 (1999): 13.

804 Edward W. Abel, "Good News From Chemistry: Meeting the Environmental Challenge," *Science in Parliament* 55 (1998): 2-3; James H. Clark, "From Brown Earth to Green Chemistry," *Science in Parliament* 55 (1998): 10. For the negative public image of chemistry, see: RSC, *Chemistry in the UK: Will it Survive?; Conclusions of the Royal Society of Chemistry Workshops for Those in Universities and Industry Held on 13/14 & 20/21 February 1995* (London: RSC, 1995), 4.

805 RSC, *The Royal Society of Chemistry: Annual Review 1998* (London: RSC, 1999), 16.

806 James H. Clark, "Editorial," *Green Chemistry* 1 (1999): G1-G2. This is Clark his inaugural editorial. See also Chapter 3.

geopolitically – in contrast to the USA and the UK – to the 'lead states' in advocating strong emission reductions, for instance within the multilateral *1979 Convention on Long-Range Transboundary Air Pollution* (*LRTAP*) regime.[807] This strict command-and-control policy symbolizes the high sociopolitical value of the environment in the Netherlands and how its administrative credibility was secured.[808] In the Dutch civic epistemology though, the policy making process was also characterized by a highly decentralized search for "consensus" and balancing of interests among many stakeholders through negotiation. The Dutch government proved reluctant to impose unilateral measures.[809] As a result, from the mid-1980s onwards several environmental covenants were negotiated. The publishing in 1988 of *Concerns for Tomorrow: A National Environmental Survey 1985-2010* [*Zorgen voor Morgen: Nationale Milieuverkenning 1985-2010*] can also be seen as an outcome of the Dutch consensus seeking political culture.[810]

This substantial report, written in response to the widely known United Nations (UN) publication of *Our Common Future*, provided the intellectual basis for new environmental policies, as well as for new rationales for Dutch chemical science in the early 1990s.[811] This led to new knowledge demands such as pollution prevention, as 'end-of-pipe sanctions were expensive', and to the replacement of fossil feedstocks by renewable ones.[812] The change in rationale came strongly to the fore in the science policy program Sustainable Technological Development [Duurzame Technologische Ontwikkeling (DTO)] (STD) initiated by the Dutch government. This programmatic alliance between science, state and industry followed, in line with the Dutch consensus seeking political culture, the foresight principle in policy making.[813] It is from this broadly conceived style of science policy making that green and sustainable

---

807 Pamela S. Chasek, David L. Downie and Janet W. Brown, *Global Environmental Politics* (Philadelphia, PA: Westview Press, 2010), 124. See also: SCP, *Sociaal en Cultureel Rapport 1998: 25 Jaar Sociale Verandering* (Rijswijk: SCP, 1998), 757-758.

808 See also Figure 6.1.

809 E.H.T.M. Nijpels (Minister for Housing, Physical Planning and Environment), "Preventie en Hergebruik van Afvalstoffen," letter to the Dutch House of Representatives, October 10, 1988. The power of the Dutch decentralized and consensus seeking style of policy making during the 1980s and 1990s was recognized by the USA and the UK. See: Henk te Velde, "Het Poldermodel. Een Introductie," in *Harmonie in Holland: Het Poldermodel van 1500 tot Nu*, ed. Dennis Bos, Maurits Ebben and Henk te Velde (Amsterdam: Uitgeverij Bert Bakker, 2011), 9-29.

810 R.B.J.C. van Noort, preface to *Zorgen voor Morgen: Nationale Milieuverkenning 1985-2010*, ed. F. Langeweg (Alphen aan de Rijn: RIVM, 1988), IX-XI.

811 World Commission on Environment and Development, *Our Common Future* (Oxford: Oxford University Press, 1987).

812 F. Langeweg, summary to *Zorgen voor Morgen*, Langeweg, XIII-XXVIII. All translations of quotations in this chapter are mine, unless otherwise stated.

813 Philip J. Vergragt and Leo Jansen, "Sustainable Technological Development: The Making of a Dutch Long-term Oriented Technology Programme," *Project Appraisal* 8 (1993): 134-140.

chemistry programmatically emerged – in times of a negative public image of chemistry, as painfully recognized by the KNCV.[814]

Chapter 4 shows that STD started in 1993 under leadership of two science policy experts with a background in chemistry: Leo Jansen and Philip Vergragt.[815] As a result of STD, one year later the program Sustainable Technological Development Chemistry [Duurzame Technologische Ontwikkeling Chemie (DTO-Chemie)] (STD-Chemistry) started, which was headed by the former KNCV President and industrial leader Jan Mulderink. Members from academia, NGOs, industry and government agencies were involved in STD-Chemistry. They identified different (applied) research areas that required additional R&D funding in order to accomplish a sustainable development. These research areas were mostly concerned with the efficient use of (preferably renewable) feedstocks as examples of sustainable chemistry. STD had two main target groups: the public and societal opinion-makers, and next to that scientists. Policymakers and scientists held the view that the public 'should regain trust in technology as a contribution to sustainability'.[816] Improving the public image of science, and chemistry in particular, was part of the effort. That's why STD-Chemistry also received recognition from Dutch chemists, who also started publishing in *Green Chemistry*.[817] In 1997 STD-Chemistry continued as the Foundation for Sustainable Chemistry Development [Stichting Duurzame Chemie Ontwikkeling]. As shown in Chapter 4, the terms "sustainable chemistry" and "green chemistry" coexisted in the 1990s, but the first appeared in the Netherlands to be more dominant than the latter. The term "sustainable chemistry" was chosen on pragmatic grounds, as it was linguistically derived from the important policy goal of "sustainable development".

In sum, green chemistry emerged in the USA, the UK and in a modified form in the Netherlands due to changing rationales for chemical science, shaped by public environmental concerns, and in response to a deteriorating image of chemistry. This emergence occurred along different historical lines and from different (institutional) origins, ranging from an academic laboratory (UK) to the policy office of an environmental controlling agency (USA). In the USA an environmental act (PPA) shaped the origins of green chemistry, which inspired policy officers (e.g., Anastas) of OPPT to disseminate the pollution prevention message

---

814 See: Barend van der Meulen and Arie Rip, "Mediation in the Dutch Science System," *Research Policy* 27 (1998): 757-758. See also Chapter 4 and: Taskforce KNCV, *Milieu: Chemici een Zorg* (Den Haag: KNCV, 1990).

815 Vergragt and Jansen, "Sustainable Technological Development," 134-140.

816 Ibid., 137-139.

817 C. Okkerse and H. van Bekkum, "From Fossil to Green," *Green Chemistry* 1 (1999): 107-114.

among chemists in order to prevent the production of (toxic) waste. This happened within the environmental chemistry division (DEC) of ACS especially during the birth of green chemistry. The Presidential Green Chemistry Challenge Awards also strengthened the institutional consolidation of green chemistry. Environmental controlling agencies or environmental chemistry divisions (ECG and MC), by contrast, did not play a significant role in the emergence of green chemistry within the UK and the Netherlands. In the UK, a pragmatic synthetic chemist heading an academic laboratory and devoted to the design of new (industrial) chemical processes with less (toxic) waste was the dominant key-actor. Clark also became the editor of the very first, and still most important, green chemistry labeled journal (*Green Chemistry*). The establishment of this RSC journal in 1999 symbolizes a turning point in the growth pattern of green chemistry. Since the start of the century the use of the term green chemistry grew enormously.[818] The objective of the journal was also to improve the public image of chemistry by stimulating chemists to publish new environmentally friendly chemical processes. Dutch chemists had similar concerns, as shown by the science policy initiative STD-Chemistry that advocated sustainable development as a broader rationale than pollution prevention only. These chemists were often concerned with the sustainable use of chemical feedstocks under the banner of sustainable chemistry, but they also published in *Green Chemistry*.[819] An interesting question that remains, is whether these three different (national) lines of emergence, also led to three different scientific types of green chemistry? This will be discussed in the next section.

6.2 An Umbrella Term: the Scientific Shaping of Green Chemistry

The previous section (6.1) shows that contextually steered recognition of green chemistry at policy levels helped to make "green chemistry" the term of choice in international scientific communication, far more than related terms such as "clean technology" or "sustainable chemistry" (Figure 6.2). The research results incorporated in Figure 6.2 also illustrate the power that green chemistry obviously had as a rhetorical mediator between chemical science and society, in comparison to related terms. As discussed in Chapter 1, some STS-scholars

---

[818] See Figure 3.2 for the share of *Green Chemistry* in the international output of green chemistry labeled articles.

[819] Nowadays, sustainable chemistry is in the Netherlands still strongly connected with the sustainable use of feedstocks, and also with the perspective to regard waste as a resource. See for example: Tom Keijer, Vincent Bakker and J. Chris Slootweg, "Circular Chemistry to Enable a Circular Economy," *Nature Chemistry* 11 (2019): 190-195.

characterize green chemistry therefore as mere rhetoric, and they saw the use of the term as an act of greenwashing, or as a buzzword, without much 'substance'.[820] But other STS-scholars were more focused on views and on scientific content in their study of green chemistry. They called the new field a "meta-discipline", resulting from an "epistemic community", or a "social or scientific movement". Chapter 5 showed that green chemistry indeed also has epistemic components. In the present section, I draw some conclusions on the (intellectual) origins of green chemistry by characterizing how chemists used the term green chemistry and I elaborate the steering role of new knowledge on the growth of green chemistry. I will argue that green chemistry can be best characterized as an 'umbrella term'.[821]

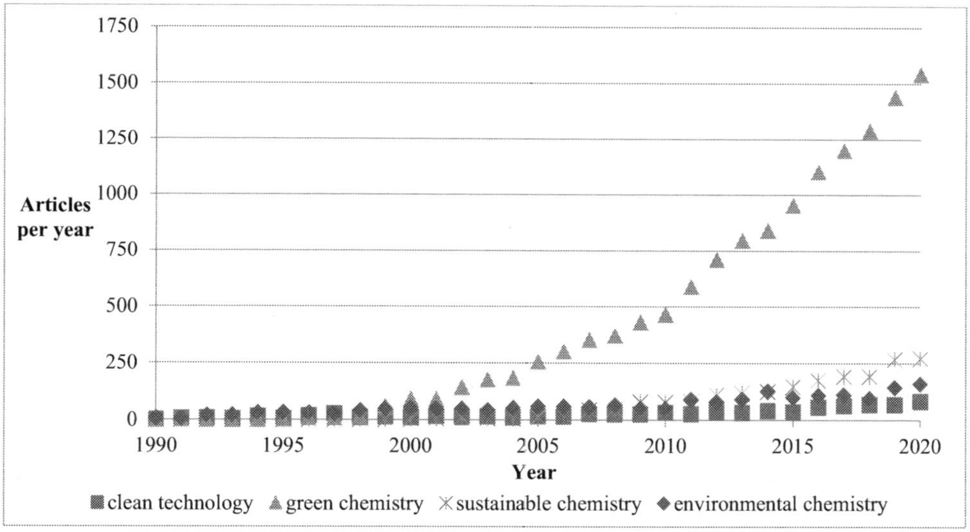

Figure 6.2: The growth of the use of the term green chemistry in scientific publications.[822]

As argued in section 6.1, the institutional origins of dominant advocates of green chemistry differed. Despite those different institutional origins, and also different rationales,

---

[820] Bernadette Bensaude-Vincent, "The Politics of Buzzwords at the Interface of Technoscience, Market and Society: the Case of 'Public Engagement in Science'," *Public Understanding of Science* 23 (2014): 240. See also: Jean-Pierre Llored, "Towards a Practical Form of Epistemology: The Case of Green Chemistry," *Studia Philosophica Estonia* 5 (2012): 42. He emphasizes the 'strong heuristic power' of green chemistry.

[821] Arie Rip and Jan-Peter Voss, "Umbrella Terms as Mediators in the Governance of Emerging Science and Technology," *Science, Technology & Innovation Studies* 9 (2013): 42.

[822] These figures are obtained from ISI Web of Knowledge: Web of Science by searching on "topic" (Accessed August 12, 2021).

many chemists published their research in *Green Chemistry* (Chapter 5), and by so doing labelled it as green chemistry. This implies that the rhetorical use of the term green chemistry was not bounded to strict rules, which matches with the concept of an 'umbrella term'.[823]

In Chapter 5 I have shown that quite different environmentally-related terms were often combined in scientific arguments. Clark, for instance, used the terms clean technology and green chemistry both in his publications, without noticeable differences in epistemic meaning.[824] And the Dutch chemists Cees Okkerse and Herman van Bekkum, who were part and parcel of the "sustainable chemistry" labeled STD-Chemistry policy program, published in 1996 their chemical views on the replacement of fossil feedstocks by renewable ones, as an example of "sustainable chemistry".[825] Three years later they published essentially the same findings in *Green Chemistry* though.[826] The mixing of (new) environmentally-related terms can be observed in many scientific publications.[827] The mixing of these environmentally-related terms had as a disadvantage though, that it blurred to some degree the internal scientific identity of green chemistry. This is illustrated by scientific controversies about the meaning of these terms.[828]

Chapter 5 also discusses other rhetorical characteristics in the use of the term green chemistry by chemists. For simplicity purposes, I categorize these chemists in two groups: passive users and active users. The first group often uses the term green chemistry in their scientific publications without (explicitly) referring to important theoretical concepts

---

823 See Chapter 1. See also: Rip and Voss, "Umbrella Terms as Mediators," 44-46. According to Rip and Voss, the institutional origins of umbrella terms can be found in politics, science policy as well as in science itself.

824 Clark published new chemical syntheses aiming at pollution prevention, which were often based on "supported reagent type catalysts" (Chapter 5). See also: Office of Science and Technology, *Progress through Partnership*, 80-81. Catalysis belonged to the 'generic science and technology priorities' within TF.

825 C. Okkerse and H. van Bekkum, "Vernieuwbare Grondstoffen voor de Chemische Industrie," in DTO-Chemie, *Duurzaamheid en Chemie: Een Bundel Essays* (Delft: DTO, 1996), 11-43.

826 Okkerse and Van Bekkum, "From Fossil to Green," 107-114.

827 See for example: David Allen, "Welcome to *ACS Sustainable Chemistry & Engineering*," *ACS Sustainable Chemistry & Engineering* 1 (2013): 1; Environment Directorate, *Need for Research and Development Programmes in Sustainable Chemistry* (Paris: OECD, 2002), 18. The Organisation for Economic Co-operation and Development (OECD) mixed the terms "sustainable chemistry" and "green chemistry", but recognized the rhetorical power of the latter with respect to improving the image of chemistry.

828 See Otto Hutzinger in Chapter 5. See also the following debate: Michael P. Wilson and Megan R. Schwarzman, "Toward a New U.S. Chemicals Policy: Rebuilding the Foundation to Advance New Science, Green Chemistry, and Environmental Health," *Environmental Health Perspectives* 117 (2009): 1202-1209; Karen P. O'Brien, John P. Meyers and John Warner, "Green Chemistry: Terminology and Principles," *Environmental Health Perspectives* 117 (2009): A385-A386; Michael P. Wilson and Megan R. Schwarzman, "Green Chemistry: Wilson and Schwarzman Respond," *Environmental Health Perspectives* 117 (2009): A386.

developed in green chemistry. Those chemists published research results in order to show that they are committed to the tackling of social questions ranging from the minimizing of toxicity impacts of chemical processes to the replacement of fossil feedstocks or to pollution prevention.[829]

The active users, by contrast, formulated comprehensive views or definitions in order to create a conceptual identity for green chemistry, and attempted to persuade other chemists to embrace their views. This group is much smaller than the first one and has been dominated by Anastas.[830] He greatly contributed to the intellectual identity of the green chemistry field. In the USA, this field started with chemical knowledge production that was often connected with pollution prevention in order to decrease emissions of toxic wastes in particular, but based on rather different intellectual sources.[831] During the birth of green chemistry, Anastas highlighted the importance of catalysis and atom economy in his conceptualization of green chemistry. Together with John Warner he published in 1998 a more comprehensive and substantive demarcation of green chemistry: the frequently cited "12 principles" (Figure 5.2).[832] These principles form a conceptual framework that provided guidelines for chemists ranging from pollution prevention, to health considerations and to an increased use of renewable feedstocks. As discussed in Chapter 1, several STS-scholars argued that these principles had a major intellectual impact on this growing field. But, as Chapter 5 shows, this conclusion is open to debate, because these 12 guidelines for working in the laboratory have a "should be sense" rather than a "must be sense".[833]

Chapter 5 also shows that in the Netherlands the use of the terms sustainable and green chemistry was largely rhetorical during the 1990s and not characterized by a well-defined

---

[829] Rip and Voss, "Umbrella Terms as Mediators," 39-59; Bensaude-Vincent, "The Politics of Buzzwords," 238-253. See also: Laura Maxim, "More than a Scientific Movement: Socio-political Influences on Green Chemistry Research in the United States and France," *Science & Technology Studies* 31 (2018): 33 and 37. My characterization of passive users and active users is in line with the findings of Maxim.

[830] Based on searching the topic "green chemistry", ISI Web of Knowledge: Web of Science lists approximately 14,500 articles with each a minimum of one author (Accessed August 12, 2021). About 500 authors used the term green chemistry at least 8 times in their articles. Thus, a much bigger group used the term as passive users. Anastas has been the most productive writer and created a historical narrative to persuade other chemists to join green chemistry. See also: Johan A. Linthorst, "An Overview: Origins and Development of Green Chemistry," *Foundations of Chemistry* 12 (2010): 55-68.

[831] See section 5.1.1.

[832] Paul T. Anastas and John C. Warner, *Green Chemistry: Theory and Practice* (New York, NY: Oxford University Press, 1998). According to the search engine Scopus (Accessed May 22, 2022), the citation frequency was in the first years pretty low, but increased significantly in the late 2000s.

[833] See Chapter 5.

concept or paradigm. The terms were mostly used in the context of renewable (bio)feedstocks in order to replace fossil feedstocks[834], and also with respect to pollution prevention, "dematerialized energy" and a reduced and efficient use of chemical feedstocks in general. In the UK, by contrast, the "active user" Clark formulated his 'ideal synthesis' concept.[835] This should be seen as a pragmatic approach to the design of chemical syntheses, which recognizes contextually steered rationales such as pollution prevention, economic targets and safety considerations. He therefore aimed for 100 % "atom efficiency". This matched with the concept of atom economy, which could be applied to objects traditionally studied by chemists and which could help to bridge the laboratory with societal demands.[836]

Obviously, the intellectual origins and content of green chemistry differ between the three countries investigated, as witnessed by substantial differences in the chemical knowledge connected with this term (Chapter 5). The emergence of this field was not centered around a worldwide central research problem.[837] Therefore, I conclude that the growth of green chemistry is not following the development path of a traditional scientific specialty, as put forward by Nicolas Mullins, which would start with a small set of scientists in the laboratory that was jointly involved in solving the same well-defined research problems.[838] In fact, chemists who were working on a wide variety of research problems and who built on rather different intellectual sources published in *Green Chemistry* under the umbrella of green chemistry.[839]

In summary, I conclude that the scientific content and scope of green chemistry are highly varied due to different intellectual origins and different policy objectives. The research problems and aims associated with the term green chemistry varied too, ranging from

---

[834] The term green chemistry was often associated with renewable feedstocks, especially in the early 1990s. See also Chapter 4.
[835] James H. Clark, "Green Chemistry: Challenges and Opportunities," *Green Chemistry* 1 (1999): 2. This article has been frequently cited. See Chapter 5.
[836] Ibid. Clark did not cite in his article the work of Barry Trost though, the man who introduced the concept of atom economy.
[837] For example, *Green Chemistry* published special issues from rather different knowledge areas, e.g., renewable resources (April 1999) and sonochemistry (June 1999).
[838] Nicholas C. Mullins, "The Development of a Scientific Specialty: the Phage Group and the Origins of Molecular Biology," *Minerva* 10 (1972): 51-82. See also Figure 3.2. The huge number of journals, which publishes green chemistry labeled articles, is very uncommon to a scientific specialty. Scientists in the same specialty typically concern about six journals. See: K. Brad Wray, "Rethinking the Size of Scientific Specialties: Correcting Price's Estimate," *Scientometrics* 83 (2010): 471-476. The claim of Wray is based on scientometric analysis.
[839] See Figure 5.4 for the weak bibliographical coupling. See also: Carlos Alberto Marques and Adélio A.S.C. Machado, "An Integrated Vision of the Green Chemistry Evolution along 25 Years," *Foundations of Chemistry* 23 (2021): 323.

renewable feedstocks (the Netherlands) to pollution prevention (the USA), especially during the emergence of the field. All those different aspects of green chemistry shaped a heterogeneous intellectual identity and found a platform though under the umbrella of the journal *Green Chemistry*. The launching of this journal gave the growth of the field a real boost, which was further spurred by the "user-friendliness" of the green chemistry conceptualizations by Clark (ideal synthesis) and Anastas and Warner (the 12 principles), and the "user-friendliness" of the term itself, as well as its rhetorical strength. Chemists from rather different institutional and intellectual backgrounds who were willing to tackle environmental problems and to present their results to other chemists and to society recognized the umbrella term green chemistry.

6.3 Conclusions and Outlook: the Meaning of Green Chemistry

What has "green chemistry" to offer for chemists and for society, at the end of the day? This thesis, *Research between Science, Society and Politics: The History and Scientific Development of Green Chemistry*, has shown that many chemists, from rather different origins and different political cultures, recognized green chemistry as a new field and were eager to produce new chemical knowledge in order to tackle environmental problems. This thesis also shows that green chemistry emerged (institutionally) as a rhetorical response by chemical societies to the negative public image of chemistry. The term green chemistry has been considered by chemists from the start as 'meaningful to the general public'.[840] The diversity in institutional origins and epistemic components of green chemistry, combined with its rhetorical "user-friendliness", characterize green chemistry as an umbrella term.

The term green chemistry also has been, and is, a mediator at the cutting-edge of science, society and politics at various occasions, the Nobel Prize in Chemistry included.[841] The widespread use was spurred by the "user-friendliness" as stated in the previous section (6.2), but led meanwhile to different beliefs and justifications concerning the value and scope

---

[840] Roger Sheldon, "Green Chemistry – One Year On," *Green Chemistry* 2 (2000): G1-G4. Though, in Germany there was also opposition to the term green chemistry during the 1990s due to socio-political feelings (see section 5.3).
[841] "The Nobel Prize in Chemistry 2005: Yves Chauvin, Robert H. Grubbs, Richard R. Schrock," The Royal Swedish Academy of Sciences, accessed May 22, 2022, http://www.nobelprize.org/nobel_prizes/chemistry/laureates/2005/press.html. See: Par Jean-Louis Hérisson and Yves Chauvin, "Catalyse de Transformation des Oléfines par les Complexes du Tungstène. II. Télomérisation des Oléfines Cycliques en Présence d'Oléfines Acycliques," *Makromolekulare Chemie* 141 (1971): 161-176. This publication is in advance of the birth of green chemistry, which demonstrates once again the "user-friendliness" of this term.

of green chemistry. Attempts to assess the greenness of chemical knowledge should be seen in that light.[842] Characteristic of an umbrella term, the scientific meaning and epistemic status of green chemistry have been contested many times.[843] So, the same vagueness and scientific broadness that contributed to the successful growth of the term green chemistry, also make it vulnerable. It demonstrates the ambivalence in the scholarly recognition of green chemistry.[844]

Since the birth of the field, green chemists have not been centered around a unified paradigm or scientific theory. Chapter 5 shows that the demarcation offered by the 12 principles has been weak. In contrast to for instance Marianna Epicoco et al. ("epistemic community") and Jody Roberts ("scientific movement"), I conclude that the twelve principles of green chemistry lack the disciplinary power to create intellectual coherence.[845] Despite the spectacular growth of green chemistry, the emergence of this new field has not been a "scientific revolution" in the Kuhnian sense.[846]

Finally, this book shows that the umbrella term green chemistry has the power to set research agendas in order to tackle challenging knowledge demands for the benefit of man, society and the environment. It also shows that green chemists have been working in rather different research areas.[847] Understandably, the meaning and epistemic value of the scientific knowledge they produce is hard to understand for society due to limited scientific literacy. As outlined in this thesis, this gives way to mere rhetorical responses especially in times of

---

[842] See Figure 5.5. See also: Katharine Sanderson, "Chemistry: It's not Easy Being Green," *Nature* 469 (2011): 18-20; Yvonne van der Meer, *How Sustainable Is It? That Is the Question* (Maastricht: Maastricht University, 2021); Keijer, Bakker and Slootweg, "Circular Chemistry," 192-193.

[843] See Neil Winterton in Chapter 5. See also: Colin Raston (Editorial Board Chair 2002-2005 of *Green Chemistry*) in: James H. Clark et al., "15 Years of *Green Chemistry*," *Green Chemistry* 16 (2014): 20. They argued that 'we need to be more clear on what deserves to be classified as green chemistry'. Gabrielle Fekete et al., *EPA's Presidential Green Chemistry Challenge Awards Program Lacks Adequate Support and Transparency and Should Be Assessed for Continuation* (Washington, DC: EPA Office of Inspector General, 2015), 6.

[844] See Chapter 2 and: Kira J.M. Matus et al., "Barriers to the Implementation of Green Chemistry in the United States," *Environmental Science & Technology* 46 (2012): 10892-10899. The ACS did not establish a green chemistry labeled journal during the birth of the field.

[845] See Chapter 5 for further discussion. See also: Jody A. Roberts, "Creating Green Chemistry: Discursive Strategies of a Scientific Movement," (PhD diss., Virginia Polytechnic Institute and State University, 2005), 72-74. In VirginiaTechWorks; available from https://vtechworks.lib.vt.edu/handle/10919/27529, accessed May 22, 2022; Marianna Epicoco, Vanessa Oltra and Maïder Saint Jean, "Knowledge Dynamics and Sources of Eco-innovation: Mapping the Green Chemistry Community," *Technological Forecasting & Social Change* 81 (2014): 391.

[846] See Chapter 1.

[847] This commitment is in line with the operational realism concept of Bensaude-Vincent and Simon as discussed in Chapter 1. See: Bernadette Bensaude-Vincent and Jonathan Simon, *Chemistry: The Impure Science* (London: Imperial College Press, 2008), 202-207.

increasing democratization of chemical science. This book shows that several chemists are good-willing though and have been producing valuable chemical knowledge to tackle environmental problems.

6.4 Reflection on Methodology and Opportunities for Further Research

An important pillar of the research methodology in this thesis has been its focus on the ACS, RSC and KNCV. The selection of these professional and learned bodies yielded significant results, but also posed limitations. On one hand, the impact of major institutionalizations (within ACS and RSC) on the emergence and growth of publications on green chemistry has been explained. The Chapters 2 through 4 identify and highlight so-called "active users" in this regard. On the other hand, it can be questioned how representative the ACS, RSC and KNCV are for the large group of "passive users" though, which have been active in different contexts ranging from chemical industry to chemistry education. These contexts received some attention in this thesis, but further research could improve this in order to better understand the impact of these interest areas on the growth of green chemistry.

The Chapters 2 through 4 show the pathways of emergence of green chemistry in the USA, the UK and the Netherlands, which developed along different historical lines and were conditioned with respect to the civic epistemologies in these countries. But, it is inevitably that the selection of these Western democracies limits our understanding about the emergence of green chemistry in rather different political cultures. Further research could therefore be focused on other countries as well such as African countries or Germany (see section 5.3).[848] This might enrich our historical understanding concerning the emergence of green chemistry.

I paid much attention to the intellectual origins of green chemistry through the lens of the journal *Green Chemistry*, including the bibliographic coupling among the authors who published in 1999 in this journal. The same type of analysis could also be done for later years, and perhaps also for other relevant journals, in order to determine changes of bibliographic coupling in the course of time.

It goes without saying that activities in green chemistry are often presented in the spirit of sustainable development.[849] The latter has been a policy objective of many countries since

---

[848] See for example: Pietro Tundo and Liliana Mammino, ed., *Green Chemistry in Africa* (Venice: IUPAC and INCA, 2002); Nigist Asfaw et al., "The 13 Principles of Green Chemistry and Engineering for a Greener Africa," *Green Chemistry* 13 (2011): 1059-1060; Robin D. Rogers, "First China-USA Green Chemistry Workshop," *Green Chemistry* 8 (2006): 126-127.
[849] See Chapter 5.

the late 1980s, which received geopolitical attention as well.[850] Recently, for instance, climate change received major geopolitical attention as witnessed by the 2015 Paris Climate Change Conference of the UN. Perhaps we are therefore at the start of a new era that impacts science policies with new knowledge demands, which are mediated by international organizations, e.g., the UN and the OECD. The role of bodies such as these, with respect to the rationale for the chemical sciences, has been underexposed in this thesis, but I assume they are also relevant for the understanding of mechanisms impacting on the chemical sciences with respect to the development of a greener chemistry.

---

[850] World Commission on Environment and Development, *Our Common Future*.

Sources and Bibliography

(A) Interviews and Personal Communications

James H. Clark, (phone) interview, December 12, 2010.

Jurgen Exner, (phone) interview, July 23, 2010.

Carol A. Farris, interview, Washington, DC, July 26, 2010.

Ray Garant, interview, Washington, DC, July 22, 2010.

Jan J.M. Mulderink, interview, Rozendaal (NL), August 30, 2011.

Dick Sijm, interview, Bilthoven (NL), August 24, 2011.

Jennifer Young, interview, Washington, DC, July 23, 2010.

(B) Archives and Collections

* The Netherlands

KNCV, Leidschendam

Stuurgroep Sectie Milieuchemie, *Koninklijke Nederlandse Chemische Vereniging: Sectie Milieuchemie*, August 4, 1970.

"KNCV Jaarverslag," published annually in the *Chemisch Weekblad*, 1972-1994.

[Pierre del Castilho, a former board member of the KNCV Division of Environmental Chemistry [Sectie Milieuchemie] (MC), gave me access to:]

Jaarboek: Sectie Milieuchemie (Den Haag: KNCV), 1990-2005.

Minutes Executive Committee Meeting of MC, 1990-1995.

Staten-Generaal Digitaal 1814-1995, www.statengeneraaldigitaal.nl

J.G.M. Alders and J.R.H. Maij-Weggen, "Nationaal Milieubeleidsplan," letter to the members of the Dutch House of Representatives, May 28, 1990.

J.E. Andriessen, "Technologie en Milieu," letter to the members of the Dutch House of Representatives, April 1, 1992.

Ministerie van Volksgezondheid & Milieuhygiëne, "Kamerstuknummer 11906, ondernummer 2: Urgentienota Milieuhygiëne," July 4, 1972.

E.H.T.M. Nijpels, "Preventie en Hergebruik van Afvalstoffen," letter to the Dutch House of Representatives, October 10, 1988.

Ed H.T.M. Nijpels, "Rijksbegroting voor het Jaar 1989," letter to the Dutch House of Representatives, April 10, 1989.

E.H.T.M. Nijpels, "Nationaal Milieubeleidsplan (NMP)," letter to the Dutch House of Representatives, May 25, 1989.

J.M.M. Ritzen, "Wetenschapsbudget 1991: Kamerstuk 21805," letter to the Dutch House of Representatives, September 18, 1990.

L.B.J. Stuyt, "Urgentienota Milieuhygiëne," letter to the members of the Dutch House of Representatives, July 4, 1972.

Tweede Kamer Commissie voor Milieubeheer, "Klimaatverandering door CO_2 en Andere Sporegassen: Lijst van Vragen," October 6, 1987.

Tweede Kamer der Staten-Generaal, "Kamerstuk 6100 XII, ondernummer 22: Verslag van het Mondeling Overleg Betreffende Afdeling VII," 1961.

Tweede Kamer der Staten-Generaal, "Kamerstuk 17555, ondernummer 7: Kabinetsformatie 1982," September 13, 1982.

Tweede Kamer der Staten-Generaal, "Kamerstuk 18605, ondernummer 2: Indicatief Meerjaren Programma Lucht 1985-1989," September 18, 1984.

Tweede Kamer der Staten-Generaal, "Kamerstuk 21137: Nationaal Milieubeleidsplan (NMP)," May 25, 1989.

VNCI, Leidschendam

VNCI Jaarverslag (Den Haag: VNCI), 1980-1998.

*** The United Kingdom**

RSC, London

The Chemical Society: Annual Report of Council and Accounts (London: CS), 1970-1979.

RIC, *Royal Institute of Chemistry: Report of the Council for the Year Ended 30 September of 1971* (London: RIC, 1971).

RSC, *Chemistry in the UK: Will it Survive?; Conclusions of the Royal Society of Chemistry Workshops for Those in Universities and Industry Held on 13/14 & 20/21 February 1995* (London: RSC, 1995).

The Royal Society of Chemistry: Annual Review (London: RSC), 1986-2001.

*** The United States of America**

EPA, Washington, DC

[The following internal EPA communications were obtained from EPA officer Carol A. Farris:]

Paul T. Anastas, "Design for the Environment: Chemical Design Project," (1993): 1-2.

Paul T. Anastas, "Design for the Environment: Chemical Design Project; Changing the Way Chemists Think about Chemicals," (AL: probably early 1993): 1-2.

Paul T. Anastas and Tracy C. Williamson, "Design for the Environment: The Green Chemistry Challenge, Quick Reference Fact Sheet," (July 1996).

EPA, "The Presidential Green Chemistry Challenge Awards Program: Listing of 1996 Award Entries," (Spring 1996).

EPA, "The President's Green Chemistry Challenge Awards Program: Nomination Package," (AL: probably Summer of 1995): 1.

ACS, Washington, DC

American Chemical Society Annual Report (Washington, DC: ACS), 1987-2006.

Science History Institute, Philadelphia, PA

Minutes of the executive and open sessions of the ACS Council Committee on Divisional Activities, April 9, 1989, Dallas, TX.

Minutes of the ACS Divisional Officers Conference/Program Coordination Conference, January 27-29, 1989, Dallas, TX.

Joseph Breen, fax to Gordon Bellen, July 26, 1994.

DEC, *ACS Division of Environmental Chemistry Secretary's Report for the Year 1988* (Washington, DC: ACS, 1989).

DEC, *ACS Division of Environmental Chemistry Secretary's Report for the Year 1989: The Environmental Division's 76th Year* (Washington, DC: ACS, 1990).

DEC, *ACS Division of Environmental Chemistry Secretary's Report for the Year 1990: The Environmental Division's 77th Year* (Washington, DC: ACS, 1991).

DEC, "Division of Water, Air, and Waste Chemistry," April 10, 1973.

Minutes of the DEC Executive Committee Meeting, 1989-1999.

Minutes of the DEC Long Range Program Planning Committee Meeting, Denver, CO, March 29, 1993.

List of DEC Officers and Executive Committee Members-1990, dated July 13, 1990.

Jurgen Exner, letter to Margaret Carey, December 2, 1996.

Larry H. Keith, letter to Bob Jolly, August 18, 1989.

Cathy A. Nelson, letter to Alan Elzerman, February 22, 1993.

Susan Turner and Margaret Carey, letter to Paul Anastas, Alan Ford and the members of CEI, members of ACS Committee on Chemistry and Public Affairs, Chair and Secretary of DEC and Chair and Co-Chair of ACS Division of Industrial and Engineering Chemistry, December 13, 1996.

(C) Periodicals Systematically Scanned

American Chemical Society Annual Report (Washington, DC: ACS), 1987-2006.

The Chemical Society: Annual Report of Council and Accounts (London: CS), 1970-1980.

EnvirofACS (Washington, DC: DEC), 1992-1997.

Environmental Chemistry Group Newsletter (London: ECG), 1995-2000.

"KNCV Jaarverslag," published annually in the *Chemisch Weekblad*, 1970-1994.

Jaarboek: Sectie Milieuchemie (Den Haag: KNCV), 1990-2005.

The Royal Society of Chemistry: Annual Review (London: RSC), 1986-2001.

VNCI Jaarverslag (Den Haag: VNCI), 1980-1998.

(D) Bibliography

A

Abbadi, A., et al., "New Ca-sequestering Materials: Based on the Oxidation of the Hydrolysis Products of Lactose," *Green Chemistry* 1 (1999): 231-235.

Abel, Edward W., "Good News From Chemistry: Meeting the Environmental Challenge," *Science in Parliament* 55 (1998): 2-3.

[ACS], "American Chemical Society: 1972 Annual Report," *Chemical & Engineering News* 51 (1973): 27.

[ACS], "200th ACS National Meeting Aug. 26-31," *Chemical & Engineering News* 68 (1991): 43-137.

[ACS], "201st National Meeting Atlanta," *Chemical & Engineering News* 69 (1991): 65-67.

[ACS], "Technical Program & Exposition. 4th Chemical Congress of North America & 202nd ACS National Meeting," *Chemical & Engineering News* 69 (1991): 53-123.

[ACS], "Technical Program & Exposition: 204th ACS National Meeting and Exposition Program, Washington, D.C., August 23-28, 1992," *Chemical & Engineering News* 70 (1992): 63.

[ACS], "Technical Program: April 5-10, 1992," *Chemical & Engineering News* 70 (1992): 66.

[ACS], "Technical Program & Exposition: 206th ACS National Meeting August 22-27," *Chemical & Engineering News* 71 (1993): 84.

[ACS], "207th ACS National Meeting Technical Program & Exposition March 13-17," *Chemical & Engineering News* 72 (1994): 29-133.

[ACS], "208th National Meeting August 21-25, 1994," *Chemical & Engineering News* 72 (1994): 53-143.

[ACS], "218th American Chemical Society National Meeting & Exposition Program. New Orleans: August 22-26, 1999," *Chemical & Engineering News* 77 (1999): 59.

[ACS], [American Institute of Chemical Engineers], [Chemical Manufacturers Association], [Council for Chemical Research] and [Synthetic Organic Chemical Manufacturers Association], *Technology Vision 2020: the U.S. Chemical Industry* (Washington, DC: ACS, 1996).

[ACS Subcommittee on Environmental Improvement] and [ACS Committee on Chemistry and Public Affairs], *Cleaning our Environment: the Chemical Basis for Action* (Washington, DC: ACS, 1969).

Adema, E.H., et al., "The Heterogeneous Formation of N_2O in Air Containing NO_2, O_3 and NH_3," *Journal of Atmospheric Chemistry* 11 (1990): 255-269.

[Advisory Council on Science and Technology], *Technology Foresight: The Identification and Promotion of Emerging and Generic Technologies* (London: HMSO Publications, 1994).

Al-Gailani, Salim, "Magic, Science and Masculinity: Marketing Toy Chemistry Sets," *Studies in History and Philosophy of Science* 40 (2009): 372-381.

Allebone, J.E., R.J. Hamilton and B. Ravenscroft, "Environmental Organic Chemistry of 2,4-Dichloro-phenoxyacetic Acid," in *Specialist Periodical Reports: Environmental Chemistry Volume 1*, ed. G. Eglinton (London: CS, 1975), 160-190.

Allen, David, "Welcome to *ACS Sustainable Chemistry & Engineering*," *ACS Sustainable Chemistry & Engineering* 1 (2013): 1.

Amato, Ivan, "The Slow Birth of Green Chemistry," *Science* 259 (1993): 1538-1541.

Anastas, Paul T., "Benign by Design Chemistry," in *Benign by Design: Alternative Synthetic Design for Pollution Prevention*, eds. Paul T. Anastas and Carol A. Farris (Washington, DC: ACS, 1994), 2-22.

Anastas, Paul T., "Joe Breen – Heart and Soul of Green Chemistry," *Green Chemistry* 1 (1999): G87.

Anastas, Paul T., et al., "Condensation of Ethyl Mandelate with a Cinnamate Ester: Bicyclocarbolactone Formation," *Journal of Chemical Research (S)* 3 (1991): 54-55.

Anastas, Paul T., et al., "Promoting Green Chemistry Initiatives: Supported by a Rapidly Growing Infrastructure, the Field Promises Innovative Solutions to Pressing Environmental Problems," *Environmental Science & Technology* 33 (1999): 116A-119A.

Anastas, Paul T., and Joseph J. Breen, "Design for the Environment and Green Chemistry: The Heart and Soul of Industrial Ecology," *Journal of Cleaner Production* 5 (1997): 97-102.

Anastas, Paul T., and Carol A. Farris, eds., *Benign by Design: Alternative Synthetic Design for Pollution Prevention* (Washington, DC: ACS, 1994).

Anastas, Paul T., and Mary M. Kirchhoff, "Origins, Current Status, and Future Challenges of Green Chemistry," *Accounts of Chemical Research* 35 (2002): 686-694.

Anastas, Paul, Mary Kirchhoff and Tracy Williamson, "More 1999 Presidential Green Chemistry Awards," *Green Chemistry* 1 (1999): G174-G175.

Anastas, Paul T., and Robert Stevenson, "Synthesis of Natural Lignan Arylnapthalene Lactones, Deurinol and Retrochinensin," *Journal of Natural Products* 54 (1991): 1687-1691.

Anastas, Paul T., and John C. Warner, *Green Chemistry: Theory and Practice* (New York, NY: Oxford University Press, 1998).

Anastas, Paul T., and Tracy C. Williamson, eds., *Green Chemistry: Designing Chemistry for the Environment* (Washington, DC: ACS, 1996).

Anastas, Paul T., and Tracy C. Williamson, "Green Chemistry: An Overview," in *Green Chemistry: Designing Chemistry for the Environment*, eds. Paul T. Anastas and Tracy C. Williamson (Washington, DC: ACS, 1996), 1-17.

Anastas, Paul T., and Julie B. Zimmerman, "Design through the 12 Principles of Green Engineering," *Environmental Science & Technology* 37 (2003): 94A-101A.

Arnold, Richard, and Andrew B. Whitford, "Organizational Dilemmas of the US EPA: Why Structures Matter for Environmental Protection," *Environmental Politics* 14 (2005): 118-123.

Asfaw, Nigist, et al., "The 13 Principles of Green Chemistry and Engineering for a Greener Africa," *Green Chemistry* 13 (2011): 1059-1060.

B

Baneke, David, "Synthetic Technocracy: Dutch Scientific Intellectuals in Science, Society and Culture, 1880-1950," *British Journal for History of Science* 44 (2011): 89-113.

Barnes, Barry, and David Edge, eds., *Science in Context: Readings in the Sociology of Science* (Milton Keynes: The Open University Press, 1982).

Bassoli, Angela, et al., "Oxidation of Phenolic Phenylpropenoids with Dioxygen Using Bis(salicylideneimino)ethylenecobalt(II)," in *Green Chemistry: Designing Chemistry for the Environment*, eds. Paul T. Anastas and Tracy C. Williamson (Washington, DC: ACS, 1996), 92-104.

Bekkum, H. van, and A. Fuchs, "De Verwachtingen Zijn Hooggespannen," *Chemisch Magazine* 6/7 (1991): 334-339.

Bennett, Graham, "The History of the Dutch National Environmental Policy Plan," *Environment* 33 (1991): 7-9, 31-33.

Bensaude-Vincent, Bernadette, "The Politics of Buzzwords at the Interface of Technoscience, Market and Society: the Case of 'Public Engagement in Science'," *Public Understanding of Science* 23 (2014): 238-253.

Bensaude-Vincent, Bernadette, and Jonathan Simon, *Chemistry: The Impure Science* (London: Imperial College Press, 2008).

Berg, J.R. van den, "Groene Chemie ook Interessant voor Shell," *Chemisch Magazine* 6/7 (1991): 322-329.

Biekart, Jan W., Anton Nigten and Dick Stoppelenburg, eds., *Duurzame Chemie: Werkstrategieën voor Duurzaamheid in de Chemische Industrie met een Toetsing aan de Huidige Praktijk* (Utrecht: Stichting Natuur en Milieu, Vereniging Milieudefensie, and Waddenvereniging, 1997).

Blok, Dirk P., et al., eds., *Algemene Geschiedenis der Nederlanden; Deel 15* (Haarlem: Fibula-Van Dishoeck, 1982).

Blokzijl, Wilfried, Michael J. Blandamer and Jan B.F.N. Engberts, "Diels-Alder Reactions in Aqueous Solutions. Enforced Hydrophobic Interactions Between Diene and Dienophile," *Journal of the American Chemical Society* 113 (1991): 4241-4246.

Blunden, S.J., L.A. Hobbs and P.J. Smith, "The Environmental Chemistry of Organotin Compounds," in *Specialist Periodical Reports: Environmental Chemistry Volume 3*, ed. H.J.M. Bowen (London: RSC, 1984), 49-77.

Boehmer-Cristiansen, Sonja A., "Emerging International Principles of Environmental Protection and Their Impact on Britain," *Environmentalist* 10 (1990): 95-113.

Boender, Kees, *Milieuprotest in Nederland: Sociologische Analyse van Milieusolidariteit onder Elites en Publiek* (Rijswijk: Sijthoff Pers, 1985).

Bonilla, Jesús P.Z., "Rhetoric, Induction, and the Free Speech Dilemma," *Philosophy of Science* 73 (2006): 175-193.

Boo, Marion de, and Peter van de Beek, "Diepgeworteld Wantrouwen," *NRC*, March 14, 1991, accessed May 22, 2022, https://www.nrc.nl/nieuws/1991/03/14/diepgeworteld-wantrouwen-6960265-a266131.

Bos, Dennis, Maurits Ebben and Henk te Velde, eds., 3th ed., *Harmonie in Holland: Het Poldermodel van 1500 tot Nu* (Amsterdam: Uitgeverij Bert Bakker, 2011).

Bosmans, Jac, and Alexander van Kessel, *Parlementaire Geschiedenis van Nederland* (Amsterdam: Boom, 2011).

Bottelberghs, P.H., "Risk Analysis and Safety Policy Developments in the Netherlands," *Journal of Hazardous Materials* 71 (2000): 59-84.

Bowen, H.J.M., ed., *Specialist Periodical Reports: Environmental Chemistry Volume 2* (London: RSC, 1982).

Bowen, H.J.M., ed., *Specialist Periodical Reports: Environmental Chemistry Volume 3* (London: RSC, 1984).

Brasser, L.J., et al., "Concise Presentation of SO_2-concentration Data Collected from a Regional Network of Volumetric SO_2 Samples," *Atmospheric Environment* 5 (1971): 505-509.

Breen, Joseph J., and Paul T. Anastas, "Design for the Environment: the Environmental Paradigm for the 21st century," *Chemtech* 24 (1994): 6-7.

Breen, Joseph J., and Michael J. Dellarco, eds., *Pollution Prevention in Industrial Processes: the Role of Process Analytical Chemistry* (Washington, DC: ACS, 1992).

Breen, Joseph J., and Michael J. Dellarco, "Pollution Prevention: The New Environmental Ethic," in *Pollution Prevention in Industrial Processes: the Role of Process Analytical Chemistry*, eds. Joseph J. Breen and Michael J. Dellarco (Washington, DC: ACS, 1992), 2-12.

Breen, Joseph J., and Philip E. Robinson, eds., *Environmental Applications of Chemometrics* (Washington, DC: ACS, 1985).

Breslow, Ronald, "The Greening of Chemistry," *Chemical & Engineering News* 74 (1996): 72.

Brindle, Ian D., "Green Chemistry – A Canadian Perspective," *Green Chemistry* 1 (1999): G155-G157.

Brock, William H., *The Fontana History of Chemistry* (London: Fontana Press, 1992).

Brode, Wallace R., "President's Message" in "American Chemical Society: 1969 Annual Report," *Chemical & Engineering News* 48 (1970): 2a-3a.

Broekhans, Bertien, *Hoe Milieukunde Geschiedenis Werd: Demarcatie van een Maatschappelijk Relevante Wetenschap 1970-2000* (Ede: Nijmegen University Press, 2003).

Bruijn, T.J.N.M. de, F.H.J.M. Coenen and K.R.D. Lulofs, "Pollution Prevention Projects in the Netherlands," *Journal of Cleaner Production* 4 (1996): 41-53.

Bruyn, J. de, "Behoefte aan Grondstoffen op Maat," *Chemisch Magazine* 6/7 (1991): 330-334.

Byrne, Adam, "The 1979 Convention on Long-Range Transboundary Air Pollution: Assessing its Effectiveness as a Multilateral Environmental Regime after 35 Years," *Transnational Environmental Law* 4 (2014): 37-67.

C

Carson, Rachel, *Silent Spring* (New York, NY: Houghton Mifflin Company, 1962).

Carter, Neil, *The Politics of the Environment: Ideas, Activism, Policy*, 2nd ed., (Cambridge: Cambridge University Press, 2010).

Cathcart, Clive, "Green Chemistry in the Emerald Isle," *Chemistry & Industry* 21 (1990): 684-687.

Cawse, P.A., "Inorganic Particulate Matter in the Atmosphere," in *Specialist Periodical Reports: Environmental Chemistry Volume 2*, ed. H.J.M. Bowen (London: RSC, 1982), 1-69.

Centi, Gabriele, and Siglinda Perathoner, "Catalysis and Sustainable (Green) Chemistry," *Catalysis Today* 77 (2003): 287-297.

Chasek, Pamela S., David L. Downie and Janet W. Brown, *Global Environmental Politics*, 5th ed., (Philadelphia, PA: Westview Press, 2010).

Christidou, Vasilia, Kostas Dimopoulos and Vasilis Koulaidis, "Constructing Social Representations of Science and Technology: the Role of Metaphors in the Press and the Popular Scientific Magazines," *Public Understanding of Science* 13 (2004): 347-362.

Clapp, Brian W., *An Environmental History of Britain since the Industrial Revolution* (London: Longman Group UK Limited, 1994).

Clarenburg, Louis A., "Penalization of the Environment Due to Stench: A Study of the Perception of Odorous Air Pollution by the Population," *Atmospheric Environment* 7 (1973): 333-351.

Clark, James H., "The Greening of Chemistry," *Chemistry in Britain* 34 (1998): 43-45.

Clark, James H., "From Brown Earth to Green Chemistry," *Science in Parliament* 55 (1998): 10.

Clark, James H., "Editorial," *Green Chemistry* 1 (1999): G1-G2.

Clark, James H., "Editorial," *Green Chemistry* 1 (1999): G29-G30.

Clark, James H., "Editorial," *Green Chemistry* 1 (1999): G117-G118.

Clark, James H., "Green Chemistry: Challenges and Opportunities," *Green Chemistry* 1 (1999): 1-8.

Clark, James H., et al., "Montmorillonite Supported Transition Metal Salts as Friedel-Crafts Alkylation Catalysts," *Journal of the Chemical Society, Chemical Communications* 18 (1989): 1353-1354.

Clark, James H., et al., "Catalytic Oxidation of Organic Substrates Using Alumina Supported Chromium and Manganese," *Journal of the Chemical Society, Chemical Communications* 18 (1989): 1355-1356.

Clark, James H., et al., "15 Years of *Green Chemistry*," *Green Chemistry* 16 (2014): 18-23.

Clark, James H., and Duncan J. Macquarrie, "Environmentally Friendly Catalytic Methods," *Chemical Society Reviews* 25 (1996): 303-310.

Clark, James H., and Duncan J. Macquarrie, "Catalysis of Liquid Phase Organic Reactions Using Chemically Modified Mesoporous Inorganic Solids," *Chemical Communications* 8 (1998): 853-860.

Clark, Tom, and Andrew Dilnot, "British Fiscal Policy since 1939" in *The Cambridge Economic History of Modern Britain: Structural Change and Growth, 1939-2000*, eds. Roderick Floud and Paul Johnson (Cambridge: Cambridge University Press, 2004), 368-398.

"Clean Air Act," in *Compilation of Selected Acts within the Jurisdiction of the Committee on Commerce: Environmental Law*, [Committee on Commerce U.S. House of Representatives] (Washington, DC: US Government Printing Office, 1995), 1-451.

Clean Air Act 1956 (London: HMSO Publications, 1956). In The National Archives; available from http://www.legislation.gov.uk/ukpga/Eliz2/4-5/52/enacted, accessed May 22, 2022.

Clinton, Bill, and Al Gore, *Reinventing Environmental Regulation* (Washington, DC: National Performance Review, 1995).

Coenen, Lars, *DTO Illustratieprocessen, Wordt Vervolgd...?: Onderzoek naar Vervolgtrajecten op de Illustratieprocessen van het Interdepartementaal Onderzoekprogramma Duurzame Technologische Ontwikkeling (1993-1997)* (Eindhoven: TU/e, 2000).

Colbeck, I., and R.M. Harrison, "Tropospheric Ozone," in *Specialist Periodical Reports: Environmental Chemistry Volume 3*, ed. H.J.M. Bowen (London: RSC, 1984), 1-48.

Collins, Terrence J., "Introducing Green Chemistry in Teaching and Research," *Journal of Chemical Education* 72 (1995): 965-966.

[Committee on Commerce U.S. House of Representatives], *Compilation of Selected Acts within the Jurisdiction of the Committee on Commerce: Environmental Law* (Washington, DC: US Government Printing Office, 1995).

"Comprehensive Environmental Response, Compensation, and Liability Act of 1980," in *Compilation of Selected Acts within the Jurisdiction of the Committee on Commerce: Environmental Law*, [Committee on Commerce U.S. House of Representatives] (Washington, DC: US Government Printing Office, 1995), 857-958.

Connelly, James, and Graham Smith, *Politics and the Environment: From Theory to Practice*, 2nd ed., (London: Routledge Taylor & Francis Group, 2003).

Constable, David J.C., Alan D. Curzons and Virginia L. Cunningham, "Metrics to 'Green' Chemistry – Which Are the Best?," *Green Chemistry* 4 (2002): 521-527.

Control of Pollution Act 1974 (London: HMSO Publications, 1974). In The National Archives; available from https://www.legislation.gov.uk/ukpga/1974/40, accessed May 22, 2022.

Cook, Stuart, "Green Chemistry – Evolution or Revolution?," *Green Chemistry* 1 (1999): G138-G141.

Copius Peereboom, J.W., *Chemie, Mens en Milieu: Schadelijke Stoffen in Milieu en Voeding, een Studie over Chemische Milieuverontreiniging* (Assen: Van Gorcum, 1976).

Copius Peereboom, J.W., "General Aspects of Trace Elements and Health," *The Science of the Total Environment* 42 (1985): 1-27.

Copius Peereboom, J.W., and K. Bouwer, "Environmental Science 'Milieukunde' in the Netherlands: a Review," *The Science of the Total Environment* 129 (1993): 157-170.

Copius Peereboom-Stegema, Jenny H.J., et al., "Influence of Chronic Cd Intoxication on the Alkaline Phosphatase Activity of Liver and Kidney: Biochemical, Histochemical and Histological Investigations," *Toxicology* 14 (1979): 67-80.

Copius Peereboom, J.H.J., and J.W. Copius Peereboom, "The Intake of Cadmium in the Kempen, an Area in the South of the Netherlands," *Ecotoxicology and Environmental Safety* 18 (1989): 93-108.

[Council of the European Communities and Commission of the European Communities], *Treaty on European Union* (Luxembourg: Office for Official Publications of the European Communities, 1992).

Cramer, Jacqueline M., *De Groene Golf: Geschiedenis en Toekomst van de Milieubeweging* (Utrecht: Uitgeverij Jan van Arkel, 1989).

Cramer, J., B. de Laat and G. Straten, *Preventie van Afval en Emissies in Nederland* (Den Haag: NOTA, 1991).

Crane, Diana, "Fashion in Science: Does it Exist?," *Social Problems* 16 (1969): 433-441.

Cranwell, P.A., "Environmental Organic Chemistry of Rivers and Lakes, Both Water and Sediment," in *Specialist Periodical Reports: Environmental Chemistry Volume 1*, ed. G. Eglinton (London: CS, 1975), 22-54.

Crosby, Donald G., "Review: Environmental Chemistry: an Overview," *Environmental Toxicology and Chemistry* 1 (1982): 1-8.

Crowhurst, Georgina, and Jasmine Murphy, "UK's Environmental Watchdog Brought to Heel," *European Environmental Law Review* 15 (2006): 234-237.

Curd, Martin, and Jan A. Cover, eds., *Philosophy of Science: the Central Issues* (New York, NY: W.W. Norton & Company, 1998).

Curd, Martin, and Jan A. Cover, "Commentary," in *Philosophy of Science: the Central Issues*, eds. Martin Curd and Jan A. Cover (New York, NY: W.W. Norton & Company, 1998), 211-253.

D

Dalton, Russell J., "The Greening of the Globe? Cross-national Levels of Environmental Group Membership," *Environmental Politics* 14 (2005): 441-459.

Davis, D.D., "An Operational Graduate Program in Environmental Chemistry," *Journal of Chemical Education* 51 (1974): 775-776.

[DCMR], *Annual Report 1990* (Schiedam: DCMR, 1991).

[DCO], *Jaarverslag 2003* (Wageningen: DCO, 2004).

Deitz, Victor R., ed., *Removal of Trace Contaminants from the Air* (Washington, DC: ACS, 1975).

"Denny Hjeresen," *Green Chemistry* 5 (2003): G54-G55.

DeVito, Stephen C., "General Principles for the Design of Safer Chemicals: Toxicological Considerations for Chemists," in *Designing Safer Chemicals: Green Chemistry for Pollution Prevention*, eds. Roger L. Garrett and Stephen C. DeVito (Washington, DC: ACS, 1996), 16-59.

Dickes, Geoff, "A Brief History of the Environmental Chemistry Group," *Environmental Chemistry Group Newsletter* 1 (1995): 4-5.

Dieleman, Hans, et al., *Kiezen voor Preventie is Winnen: Naar een Preventief Milieubeleid van Bedrijf en Overheid* (Den Haag: NOTA, 1991).

Diels, Otto, and Kurt Alder, "Synthesen in der Hydroaromatischen Reihe," *Justig Liebigs Annalen der Chemie* 460 (1928): 98-122.

Dinkelman, Gerda, *Verzuring en Broeikaseffect. De Wisselwerking tussen Problemen en Oplossingen in het Nederlandse Luchtverontreinigingsbeleid (1970-1994)* (Utrecht: Uitgeverij Jan van Arkel, 1995).

Dongen, Jeroen van, *Einstein's Unification* (New York, NY: Cambridge University Press, 2010).

Draths, Karen M., and John W. Frost, "Microbial Biocatalysis: Synthesis of Adipic Acid from D-Glucose," in *Benign by Design: Alternative Synthetic Design for Pollution Prevention*, eds. Paul T. Anastas and Carol A. Farris (Washington, DC: ACS, 1994), 32-45.

Driel, Corry van, "Groene Revolutie," *C_2W* 108 (2012): 5.

[DTO], *DTO Sleutel Chemie: Zon en Biomassa: Bronnen van de Toekomst* (Den Haag: Uitgeverij ten Hagen & Stam, 1997).

[DTO-Chemie], *Duurzaamheid en Chemie: Een Bundel Essays* (Delft: DTO, 1996).

[DTO-Chemie], *Duurzame Technologische Ontwikkeling Chemie: Vijf Technologie Ontwikkelingsvelden; Onderzoek op Maat* (Den Haag: Willems & Van den Wildenberg, 1997).

[DTO-Chemie], *Reactie op de Toekomst: DTO-Chemie* (Amsterdam: Duurzame Technologische Ontwikkeling, 1997).

Dukakis, Michael S., "Environmental Politics in Post-World War II America," *Resources, Conservation and Recycling* 18 (1996): 5-9.

Dunlap, Riley E., "Public Opinion in the 1980s: Clear Consensus, Ambiguous Commitment," *Environment* 33 (1991): 10-15, 32-37.

Dunlap, Riley E., and Angela G. Mertig, eds., *American Environmentalism: the U.S. Environmental Movement, 1970-1990* (Washington, DC: Taylor & Francis New York, 1992).

Dunlap, Riley E., and Angela G. Mertig, "The Evolution of the U.S. Environmental Movement from 1970 to 1990: an Overview," in *American Environmentalism: the U.S. Environmental Movement, 1970-1990*, eds. Riley E. Dunlap and Angela G. Mertig (Washington, DC: Taylor & Francis New York, 1992), 1-10.

Dyer, Norman E., "Technology Transfer/Pollution Prevention," *Journal of Hazardous Materials* 22 (1989): 264-265.

E

Earle, Martyn J., Paul B. McCormac and Kenneth R. Seddon, "Diels-Alder Reactions in Ionic Liquids: A Safe Recyclable Alternative to Lithium Perchlorate-diethyl Ether Mixtures," *Green Chemistry* 1 (1999): 23-25.

Eglinton, G., ed., *Specialist Periodical Reports: Environmental Chemistry Volume 1* (London: CS, 1975).

Eissen, Marco, et al., "Mass Efficiency as Metric for the Effectiveness of Catalysts," *Green Chemistry* 5 (2003): G25-G27.

Emmott, Neil, and Nigel Haigh, "Integrated Pollution Prevention and Control: UK and EC Approaches and Possible Next Steps," *Journal of Environmental Law* 8 (1996): 301-311.

Environment Act 1995 (London: HMSO Publications, 1995). In The National Archives; available from http://www.legislation.gov.uk/ukpga/1995/25/section/4, accessed May 22, 2022.

[Environment Directorate], *Proceedings of the OECD Workshop on Sustainable Chemistry: Venice, 15-17 October, 1998* (Paris: OECD, 1999).

[Environment Directorate], *Need for Research and Development Programmes in Sustainable Chemistry* (Paris: OECD, 2002).

[Environment Directorate], *Sustainable Chemistry: Evidence on Innovation from Patent Data* (Paris: OECD, 2011).

Environmental Protection Act 1990 (London: HMSO Publications, 1990). In The National Archives; available from www.legislation.gov.uk/ukpga/1990/43/pdfs/ukpga_19900043_en.pdf, accessed May 22, 2022.

[EPA], *Green Lights: Third Annual Report* (Washington, DC: US Environmental Protection Agency, 1994).

Epicoco, Marianna, Vanessa Oltra and Maïder Saint Jean, "Knowledge Dynamics and Sources of Eco-innovation: Mapping the Green Chemistry Community," *Technological Forecasting & Social Change* 81 (2014): 388-402.

Epling, Gary A., and Qingxi Wang, “Preparative Reactions Using Visible Light: High Yields from Pseudoelectrochemical Transformation,” in *Benign by Design: Alternative Synthetic Design for Pollution Prevention*, eds. Paul T. Anastas and Carol A. Farris (Washington, DC: ACS, 1994), 64-75.

[European Environment Agency], *Annual Report 1999* (Copenhagen: European Environment Agency, 2000).

F

Faber, Daniel, and James O’Connor, “Environmental Politics,” in *Major Problems in American Environmental History*, ed. Carolyn Merchant (Lexington, MA: D.C. Heath and Company, 1993), 551-559.

“Facts & Figures for Chemical R&D,” *Chemical & Engineering News* 64 (1986): 32-60.

Farrington, J.W., and P.A. Meyer, “Hydrocarbons in the Marine Environment,” in *Specialist Periodical Reports: Environmental Chemistry Volume 1*, ed. G. Eglinton (London: CS, 1975), 109-136.

Farris, Carol A., Harold E. Podall and Paul T. Anastas, “Alternative Syntheses and Other Source Reduction Opportunities for Premanufacture Notification Substances at the U.S. Environmental Protection Agency,” in *Benign by Design: Alternative Synthetic Design for Pollution Prevention*, eds. Paul T. Anastas and Carol A. Farris (Washington, DC: ACS, 1994), 156-165.

Fekete, Gabrielle, et al., *EPA’s Presidential Green Chemistry Challenge Awards Program Lacks Adequate Support and Transparency and Should Be Assessed for Continuation* (Washington, DC: EPA Office of Inspector General, 2015).

Floud, Roderick, and Paul Johnson, eds., *The Cambridge Economic History of Modern Britain: Structural Change and Growth, 1939-2000* (Cambridge: Cambridge University Press, 2004).

Fortuin, J.M.H., and C. Boelhouwer, "Nieuwe Reactiemedia voor de Fijnchemie in 2040," in [DTO-Chemie], *Duurzaamheid en Chemie: Een Bundel Essays* (Delft: DTO, 1996), 45-105.

Francl, Michelle, "How to Counteract Chemophobia," *Nature Chemistry* 5 (2013): 439-440.

Frickel, Scott, and Neil Gross, "A General Theory of Scientific/Intellectual Movements," *American Sociological Review* 70 (2005): 204-232.

Fuchs, A., and H. van Bekkum, "Inuline, het Vergeten Zusje van Zetmeel," *Chemisch Magazine* 6/7 (1991): 320-321.

G

Galan, Leo de, "De Instrumentele Ontwikkeling: Analytische Chemie," in *De Geschiedenis van de Scheikunde in Nederland 3: De Ontwikkeling van de Chemie van 1945 tot het Begin van de Jaren Tachtig*, eds. Ernst Homburg and Lodewijk Palm (Delft: Delft University Press, 2004), 87-103.

Garrett, Roger L., "Pollution Prevention, Green Chemistry, and the Design of Safer Chemicals," in *Designing Safer Chemicals: Green Chemistry for Pollution Prevention*, eds. Roger L. Garrett and Stephen C. DeVito (Washington, DC: ACS, 1996), 2-15.

Garrett, Roger L., and Stephen C. DeVito, eds., *Designing Safer Chemicals: Green Chemistry for Pollution Prevention* (Washington, DC: ACS, 1996).

Gay, Hannah, "Before and After *Silent Spring*: from Chemical Pesticides to Biological Control and Integrated Pest Management-Britain, 1945-1980," *Ambix* 59 (2012): 88-108.

Georghiou, Luke, "The UK Technology Foresight Programme," *Futures* 28 (1996): 359-377.

Gillette, Robert, "Environmental Protection Agency: Chaos or Creative Tension," *Science* 173 (1971): 703-707.

Glaze, William H., "Positioning Environmental Chemistry," *Environmental Science & Technology* 23 (1989): 1173.

Godin, Benoit, and Yves Gingras, "What Is Scientific and Technological Culture and How Is it Measured? A Multidimensional Model," *Public Understanding of Science* 9 (2000): 43-58.

Goldfarb, Theodore, "Environmental Legislation," in *Major Problems in American Environmental History*, ed. Carolyn Merchant (Lexington, MA: D.C. Heath and Company, 1993), 547-551.

Gordon, Glen E., "The Role of Chemistry Departments in the Environmental Education of Chemists," *Journal of Chemical Education* 51 (1974): 769.

Graaf, Leontine A. de, and Peter Kolster, "Industrial Proteins as a Green Alternative for 'Petro' Polymers: Potentials and Limitations," *Macromolecular Symposia* 127 (1998): 51-58.

"Green Chemistry & Engineering," *The Capitol Connection* 1 (1997): 2.

"Green Is Good but it Will Cost Firms Money," *Newcastle Journal*, October 25, 1990, 13.

Gross, Richard A., David L. Kaplan and Graham Swift, eds., *Enzymes in Polymer Synthesis* (Washington, DC: ACS, 1998).

Gushee, David E., "Industrial and Engineering Chemistry," in *The American Chemical Society at 125: a Recent History, 1975-2001*, ed. Kenneth M. Reese (Washington, DC: ACS, 2002), 118-119.

H

Hamlin, Christopher, "Between Knowledge and Action: Themes in the History of Environmental Chemistry," in *Chemical Sciences in the Modern World*, ed. Seymour H. Mauskopf (Philadelphia, PA: University of Pennsylvania Press, 1993), 295-321.

Hancock, Kenneth G., and Margaret A. Cavanaugh, "Environmentally Benign Chemical Synthesis and Processing for the Economy and the Environment," in *Benign by Design: Alternative Synthetic Design for Pollution Prevention*, eds. Paul T. Anastas and Carol A. Farris (Washington, DC: ACS, 1994), 23-30.

Hanney, Steve, Mary Henkel and Dagmar von Walden Laing, "Making and Implementing Foresight Policy to Engage the Academic Community: Health and Life Scientists' Involvement in, and Response to, Development of the UK's Technology Foresight Programme," *Research Policy* 30 (2001): 12013-1219.

"Has Mrs Thatcher turned green?," *Nature* 335 (1988): 479.

Hather, Gregory J., et al., "The United States of America and Scientific Research," *Plos One* 5 (2010): 1-9.

Hays, Samuel P., "From Conservation to Environment: Environmental Politics in the United States since World War II," *Environmental Review* 6 (1982): 14-41.

Hecht, David K., "Constructing a Scientist: Expert Authority and Public Images of Rachel Carson," *Historical Studies in the Natural Sciences* 41 (2011): 277-302.

Helvoort, Ton van, and Ernst Homburg, *Een Eeuw Chemische Technologie in Nederland* (Delft: Stichting Hoogewerff-Fonds, 2021).

Hendrickson, James B., "Teaching Alternative Syntheses: The SYNGEN Program," in *Green Chemistry: Designing Chemistry for the Environment*, eds. Paul T. Anastas and Tracy C. Williamson (Washington, DC: ACS, 1996), 214-231.

Hérisson, Par Jean-Louis, and Yves Chauvin, "Catalyse de Transformation des Oléfines par les Complexes du Tungstène. II. Télomérisation des Oléfines Cycliques en Présence d'Oléfines Acycliques," *Makromolekulare Chemie* 141 (1971): 161-176.

Hessels, Laurens K., and Harro van Lente, "Practical Applications as a Source of Credibility: a Comparison of Three Fields of Dutch Academic Chemistry," *Minerva* 49 (2011): 215-240.

Hester, R.E., ed., *Industry and the Environment in Perspective* (London: RSC, 1983).

Hinderink, A.P., H.J. van der Kooi and J. de Swaan Arons, "On the Efficiency and Sustainability of the Process Industry," *Green Chemistry* 1 (1999): G176-G180.

[HMIP], *Her Majesty's Inspectorate of Pollution Annual Report 1991-92: West Division Review* (London: HMSO Publications, 1993).

[HMIP], *Annual Report 1993-94* (London: HMSO Publications, 1994).

Hoffman, Andrew J., *From Heresy to Dogma: an Institutional History of Corporate Environmentalism* (San Francisco, CA: The New Lexington Press, 1997).

Hoffmann, Dieter, "Between Autonomy and Accommodation: the German Physical Society During the Third Reich," *Physics in Perspective* 7 (2005): 293-329.

Homburg, Ernst, and Lodewijk Palm, eds., *De Geschiedenis van de Scheikunde in Nederland 3: De Ontwikkeling van de Chemie van 1945 tot het Begin van de Jaren Tachtig* (Delft: Delft University Press, 2004).

Homburg, Ernst, and Lodewijk Palm, "Grenzen aan de Groei – Groei aan de Grenzen: Enkele Ontwikkelingslijnen van de Na-oorlogse Chemie," in *De Geschiedenis van de Scheikunde in Nederland 3: De Ontwikkeling van de Chemie van 1945 tot het Begin van de Jaren Tachtig*, eds. Ernst Homburg and Lodewijk Palm (Delft: Delft University Press, 2004), 3-18.

Homburg, Ernst, and Elisabeth Vaupel, eds., *Hazardous Chemicals: Agents of Risk and Change, 1800-2000* (New York: Berghahn Books, 2019).

Homburg, Ernst, and Elisabeth Vaupel, "Conclusion," in *Hazardous Chemicals: Agents of Risk and Change, 1800-2000*, eds. Ernst Homburg and Elisabeth Vaupel (New York: Berghahn Books, 2019), 376-391.

Hudlicky, T., "Environmentally Efficient Management of Aromatic Compounds: Economic Perspectives of Biocatalytic Conversion of Aromatics to Optically Pure Synthons for the

Pharmaceutical Industry," in *Green Chemistry: Designing Chemistry for the Environment*, eds. Paul T. Anastas and Tracy C. Williamson (Washington, DC: ACS, 1996), 180-197.

Huheey, James E., "Incorporating Environmental Issues into the Inorganic Curriculum," in *Green Chemistry: Designing Chemistry for the Environment*, eds. Paul T. Anastas and Tracy C. Williamson (Washington, DC: ACS, 1996), 232-238.

Huigen, Yvonne, "Milieu, Chemici een Zorg," *Chemisch Weekblad* 87 (1991): 6.

Hutten, T.J.H.M. van, "Landbouw en Industrie Moeten Elkaar Leren Verstaan," *Chemisch Magazine* 6/7 (1991): 326-329.

Hutzinger, Otto, "The Greening of Chemistry – Is it Sustainable?," *Environmental Science and Pollution Research* 6 (1999): 123.

I

Iles, Alastair, "Greening Chemistry: Emerging Epistemic Political Tensions in California and the United States," *Public Understanding of Science* 22 (2011): 460-478.

J

"Jaaroverzicht 1995," *Chemisch Weekblad* 91 (1995): 3.

Jansen, Leo, "The Challenge of Sustainable Development," *Journal of Cleaner Production* 11 (2003): 231-245.

Jasanoff, Sheila, *Designs on Nature: Science and Democracy in Europe and the United States* (Princeton, NJ: Princeton University Press, 2005).

Jastorff, Bernd, et al., "Progress in Evaluation of Risk Potential of Ionic Liquids – Basis for an Eco-design of Sustainable Products," *Green Chemistry* 7 (2005): 362-372.

Jenne, Everett A., ed., *Chemical Modeling in Aqueous Systems: Speciation, Sorption, Solubility, and Kinetics* (Washington, DC: ACS, 1979).

K

Kaplan, David L., et al., "Enzymes in Polymer Science: An Introduction," in *Enzymes in Polymer Synthesis*, eds. Richard A. Gross, David L. Kaplan and Graham Swift (Washington, DC: ACS, 1998), 2-16.

Keith, Lawrence H., et al., "Principles of Environmental Analysis," *Analytical Chemistry* 55 (1983): 2210-2218.

Keith, Larry H., and William A. Telliard, "ES&T Special Report: Priority Pollutants I. A Perspective View," *Environmental Science & Technology* 13 (1979): 416-423.

Keijer, Tom, Vincent Bakker and J. Chris Slootweg, "Circular Chemistry to Enable a Circular Economy," *Nature Chemistry* 11 (2019): 190-195.

Keijzers, Gerard, "The Evolution of Dutch Environmental Policy: The Changing Ecological Arena from 1970-2000 and Beyond," *Journal of Cleaner Production* 8 (2000): 179-200.

Keller, Morton, *America's Three Regimes: a New Political History* (New York, NY: Oxford University Press, 2007).

Kidwai, M., and R. Mohan, "Green Chemistry: an Innovative Technology," *Foundations of Chemistry* 7 (2005): 269-287.

"KNCV-Zomercongres 1995: Natuur, Bron voor Chemie," *Chemisch Weekblad* 91 (1995): 5.

Koenis, Sjaak, and Karin Bijsterveldt, "Conceptual Analysis: an Introduction" (unpublished document, Maastricht University).

Komiya, Kyosuke, et al., "New Process for Producing Polycarbonate Without Phosgene and Methylene Chloride," in *Green Chemistry: Designing Chemistry for the Environment*, eds. Paul T. Anastas and Tracy C. Williamson (Washington, DC: ACS, 1996), 20-32.

Kraus, George A., Masayuki Kirihara and Yusheng Wu, "A Photochemical Alternative to the Friedel-Crafts Reaction," in *Benign by Design: Alternative Synthetic Design for Pollution Prevention*, eds. Paul T. Anastas and Carol A. Farris (Washington, DC: ACS, 1994), 76-83.

"Kritiek Kamer Tijdens Begrotingsdebat: Ginjaar te Langzaam met Milieuwetgeving," *Nieuwe Leidsche Courant*, November 1, 1979, 3.

Kuhn, Thomas S., ed., *The Essential Tension: Selected Studies in Scientific Tradition and Change* (London: The University of Chicago Press, 1977).

Kuhn, Thomas S., "Objectivity, Value Judgement, and Theory Choice," in *The Essential Tension: Selected Studies in Scientific Tradition and Change*, ed. Thomas S. Kuhn (London: The University of Chicago Press, 1977), 320-339.

Kuhn, Thomas S., *The Structure of Scientific Revolutions*, 3th ed., (London: The University of Chicago Press, 1996).

L

Lange, Lydia L., and P.A. Frensch, "Gaining Scientific Recognition by Position: Does Editorship Increase Citation Rates?," *Scientometrics* 44 (1999): 459-486.

Langeweg, F., ed., *Zorgen voor Morgen: Nationale Milieuverkenning 1985-2010* (Alphen aan de Rijn: RIVM, 1988).

Laszlo, Pierre, "On the Self-image of Chemists, 1950-2000," *Hyle* 12 (2006): 99-130.

Latour, Bruno, *Science in Action* (Cambridge, MA: Harvard University Press, 1987).

Latour, Bruno, and Steve Woolgar, “The Cycle of Credibility,” in *Science in Context: Readings in the Sociology of Science*, eds. Barry Barnes and David Edge (Milton Keynes: The Open University Press, 1982), 35-43.

Latour, Bruno, and Steve Woolgar, *Laboratory Life: the Construction of Scientific Facts*, 2nd ed., (Princeton, NJ: Princeton University Press, 1986).

Lee, G. Fred, “Graduate Education in Environmental Chemistry,” *Journal of Chemical Education* 51 (1974): 772-774.

Lee, Stephen J., *Aspects of British Political History 1914-1995* (London: Routledge, 1996).

Leitner, Walter, “Toward Benign Ends,” *Science* 284 (1999): 1780-1781.

Leitner, Walter, “Green Chemistry: a New Phase,” *Green Chemistry* 6 (2004): G1.

Lester, Tim, “Clean Technology Programme,” *Green Chemistry* 1 (1999): G69-G71.

Levere, Trevor H., *Transforming Matter: a History of Chemistry from Alchemy to the Buckyball* (Baltimore, MD: The Johns Hopkins University Press, 2001).

Li, Shuang, “简论绿色化学的起源与发展,” *Science & Culture Review* 14 (2017): 44-59.

Linthorst, Johan A., “An Overview: Origins and Development of Green Chemistry,” *Foundations of Chemistry* 12 (2010): 55-68.

Linthorst, Johan A., “The Image of Chemistry and Curriculum Changes,” *Educacion Quimica* 23 (2012): 240-242.

Llored, Jean-Pierre, “Towards a Practical Form of Epistemology: The Case of Green Chemistry,” *Studia Philosophica Estonia* 5 (2012): 36-60.

Llored, Jean-Pierre, and Stéphane Sarrade, "Connecting the Philosophy of Chemistry, Green Chemistry and Moral Philosophy," *Foundations of Chemistry* 18 (2016): 125-152.

Long, Bill L., *International Environmental Issues and the OECD 1950-2000* (Paris: OECD, 2000).

Lynch, William T., "Second-guessing Scientists and Engineers: Post Hoc Criticism and the Reform of Practice in Green Chemistry and Engineering," *Science and Engineering Ethics* 21 (2015): 1217-1240.

M

MacDougall, Daniel, et al., "Guidelines for Data Acquisition and Data Quality Evaluation in Environmental Chemistry," *Analytical Chemistry* 52 (1980): 2242-2249.

Manahan, Stanley E., *Environmental Chemistry* (Boston, MA: Willard Grant Press, 1972).

Manahan, Stanley E., *Environmental Chemistry*, 6th ed., (London: Lewis Publishers, 1994).

Marcelino, Leonardo Victor, Adilson Luiz Pinto, and Carlos Alberto Marques, "Intellectual Authorities and Hubs of Green Chemistry," *Transinformação* 32 (2020): e200031.

Marques, Carlos A., and Adélio A.S.C. Machado, "Environmental Sustainability: Implications and Limitations to Green Chemistry," *Foundations of Chemistry* 16 (2014): 125-147.

Marques, Carlos Alberto, and Adélio A.S.C. Machado, "An Integrated Vision of the Green Chemistry Evolution along 25 Years," *Foundations of Chemistry* 23 (2021): 299-328.

Matus, Kira J.M., "The ACS Green Chemistry Institute®: a Case Study of Partnerships to Promote Sustainability in the Chemical Enterprise," in *Enhancing the Effectiveness of Sustainability Partnerships*, ed. Derek Vollmer (Washington, DC: The National Academies Press, 2009), 263-292, also accessed May 22, 2022, www.nap.edu/html/12541/12541_casestudies_app.pdf.

Matus, Kira J.M., et al., "Barriers to the Implementation of Green Chemistry in the United States," *Environmental Science & Technology* 46 (2012): 10892-10899.

Mauskopf, Seymour H., ed., *Chemical Sciences in the Modern World* (Philadelphia, PA: University of Pennsylvania Press, 1993).

Maxim, Laura, "Chemists' Responsibility for the Health Impacts of Chemicals: Green Chemistry and Its Relation to Toxicology," *Hyle* 23 (2017): 61-80.

Maxim, Laura, "More than a Scientific Movement: Socio-political Influences on Green Chemistry Research in the United States and France," *Science & Technology Studies* 31 (2018): 24-46.

McLachlan, Michael, et al., "Obituary Otto Hutzinger," *Environmental Sciences Europe* 25 (2013): 28.

McMullin, Ernan, "Rationality and Paradigm Change in Science," in *Philosophy of Science: the Central Issues*, eds. Martin Curd and Jan A. Cover (New York, NY: W.W. Norton & Company, 1998), 119-138.

Meer, Yvonne van der, "How Sustainable Is It? That Is the Question," (Inaugural speech, Maastricht University, 2021), 1-20. In UM Research Publications; available from https://cris.maastrichtuniversity.nl/en/publications/how-sustainable-is-it-that-is-the-question, accessed May 22, 2022.

Meester, M.A.M., ed., *Milieuchemie. I. Onderzoeksmethoden in de Milieuchemie* (Heerlen: Open Universiteit, 2000).

Merchant, Carolyn, ed., *Major Problems in American Environmental History* (Lexington, MA: D.C. Heath and Company, 1993).

"Mercury Under Pressure," *Green Chemistry* 1 (1999): G4.

Meulen, Barend van der, "The Impact of Foresight on Environmental Science and Technology Policy in the Netherlands," *Futures* 31 (1999): 7-23.

Meulen, Barend van der, and Arie Rip, "Mediation in the Dutch Science System," *Research Policy* 27 (1998): 757-769.

Minnesma, M.E., "Dutch Climate Policy: A Victim of Economic Growth?," *Climate Policy* 3 (2003): 45-56.

Mitchell, Robert C., Angela G. Mertig and Riley E. Dunlap, "Twenty Years of Environmental Mobilization: Trends Among National Environmental Organizations," in *American Environmentalism: the U.S. Environmental Movement, 1970-1990*, eds. Riley E. Dunlap and Angela G. Mertig (Washington, DC: Taylor & Francis New York, 1992), 11-26.

Mitzschke, Andreas, "Elusive Publics: Understanding Techno-scientific Controversy and Democratic Governance in the GM Crops Debate," (PhD diss., Maastricht University, 2018), 225-230. In UM Research Publications; available from https://cris.maastrichtuniversity.nl/en/publications/elusive-publics-understanding-techno-scientific-controversy-and-d, accessed May 22, 2022.

Mol, Arthur P.J., *The Refinement of Production: Ecological Modernization Theory and the Chemical Industry* (Utrecht: Van Arkel, 1995).

Moore, John W., and Elizabeth A. Moore, "Resources in Environmental Chemistry: I. an Annotated Bibliography of the Chemistry of Pollution and Resources," *Journal of Chemical Education* 53 (1976): 167-171.

Morais, Carla, "Storytelling with Chemistry and Related Hands-on activities: Informal Learning Experiences to Prevent "Chemophobia" and Promote Young Children's Scientific Literacy," *Journal of Chemical Education* 92 (2015): 58-65.

Morris, R.J., and F. Culkin, "Environmental Organic Chemistry of Oceans, Fjords, and Anoxic Basins," in *Specialist Periodical Reports: Environmental Chemistry Volume 1*, ed. G. Eglinton (London: CS, 1975), 81-108.

Mulderink, J.J.M., "KNCV-Zomercongres 1995: Natuur, Bron voor Chemie," *Chemisch Weekblad* 91 (1995): 5.

Mulderink, Jan J.M., "Sustainability in the Chemical Industry by 2050: The Outlook for a Sustainable Chemical Industry in 2050," *Studies in Environmental Science* 72 (1998): 909-913.

Mullins, Nicholas C., "The Development of a Scientific Specialty: the Phage Group and the Origins of Molecular Biology," *Minerva* 10 (1972): 51-82.

Murphy, Mark A., "Early Industrial Roots of Green Chemistry and the History of the BHC Ibuprofen Process Invention and its Quality Connection," *Foundations of Chemistry* 20 (2018): 121-165.

N

"The National Environmental Policy Act of 1969," in *Major Problems in American Environmental History*, ed. Carolyn Merchant (Lexington, MA: D.C. Heath and Company, 1993), 498-500.

[National Research Council], *Linking Science and Technology to Society's Environmental Goals* (Washington, DC: National Academy Press, 1996).

Nelissen, Nico, Jan van der Straaten and Leon Klinkers, eds., *Classics in Environmental Studies: An Overview of Classic Texts in Environmental Studies* (Utrecht: International Books, 1997).

Nelson, Andrew, et al., "Do Innovation Measures Actually Measure Innovation? Obliteration, Symbolic Adoption, and Other Finicky Challenges in Tracking Innovation Diffusion," *Research Policy* 43 (2014): 927-940.

"New Industrial Processes," *Green Chemistry* 1 (1999): G31-G32.

"New Investments in the Czech Republic in Refinery," *Green Chemistry* 1 (1999): G3.

Nicolau, K.C., and E.J. Sorensen, *Classics in Total Synthesis: Targets, Strategies, Methods* (New York, NY: VCH, 1996).

Norton, Mary B., et al., *A People and a Nation: a History of the United States*, 3th ed., (Boston, MA: Houghton Mifflin Company, 1991).

Notten, Frank, *De Nederlandse Economie: De Invloed van de Aardgaswinning op de Nederlandse Economie* (Den Haag: Centraal Bureau voor de Statistiek, 2015).

[NWO Chemische Wetenschappen], *Jaarboek 2005* (Den Haag: NWO, 2005).

O

O'Brien, Karen P., John P. Meyers and John Warner, "Green Chemistry: Terminology and Principles," *Environmental Health Perspectives* 117 (2009): A385-A386.

[Office of Science and Technology], *Progress through Partnership: Report from the Steering Group of the Technology Foresight Programme 1995* (London: HMSO Publications, 1995).

Okkerse, C., and H. van Bekkum, "Vernieuwbare Grondstoffen voor de Chemische Industrie," in [DTO-Chemie], *Duurzaamheid en Chemie: Een Bundel Essays* (Delft: DTO, 1996), 11-43.

Okkerse, C., and H. van Bekkum, "From Fossil to Green," *Green Chemistry* 1 (1999): 107-114.

O'Mahony, Mary, "Employment, Education and Human Capital," in *The Cambridge Economic History of Modern Britain*, eds. Roderick Floud and Paul Johnson (Cambridge: Cambridge University Press, 2004), 112-133.

Opstal, Marian van, "Addertjes onder het Groen," *Chemisch Magazine* 6/7 (1991).

O'Riordan, Timothy, "The Politics of Environmental Regulation in Great Britain," *Environment* 30 (1988): 4-9.

O'Riordan, T., "Industrial Pollution Control in the UK," *The Science of the Total Environment* 129 (1993): 39-53.

Osborn, Derek, "Some Reflections on UK Environment Policy, 1970-1995," *Journal of Environmental Law* 9 (1997): 3-22.

Owens, Susan, "Experts and the Environment – the UK Royal Commission on Environmental Pollution 1970-2011," *Journal of Environmental Law* 24 (2012): 1-22.

P

Pattee, Harold E., "ACS Divisions," in *The American Chemical Society at 125: a Recent History, 1975-2001*, ed. Kenneth M. Reese (Washington, DC: ACS, 2002), 77-81.

Paul, R.C., "An Industrialist's View," in *Industry and the Environment in Perspective*, ed. R.E. Hester (London: RSC, 1983), 12-29.

Penna, Anthony N., *The Human Footprint: A Global Environmental History* (Chichester: Wiley-Blackwell, 2010).

Perrée, Harry, "Jan Willem Copius Peereboom: 'Ik Geloof in de Industrie, maar ze Moet zich Netjes Gedragen'," *Natuur en Milieu* 21 (1997): 14-15.

"Pollution Prevention Act of 1990," in *Compilation of Selected Acts within the Jurisdiction of the Committee on Commerce: Environmental Law*, [Committee on Commerce U.S. House of Representatives] (Washington, DC: US Government Printing Office, 1995), 615-623.

Porter, George, "The Amalgamated Societies," *Nature* 234 (1971): 511-512.

Pratt, George L.B., "Educational Needs of Chemists in Pollution Control and Abatement," *Journal of Chemical Education* 51 (1974): 770-771.

Price, David, "Sink or Swim?: Ozone Depletion, Friends of the Earth, Consumers' Association, and European Campaign Against Cancer," *Energy Policy* 17 (1989): 295-296.

Primio, Juan C. di, "Data Quality and Compliance Control in the European Air Pollution Regime," in *The Implementation and Effectiveness of International Environmental Commitments: Theory and Practice*, eds. David G. Victor, et al. (Cambridge, MA: the MIT Press, 1998), 283-304.

Pritchard, Duncan, "Recent Work on Epistemic Value," *American Philosophical Quarterly* 44 (2007): 85-110.

Pugh, Martin, *State & Society: A Social and Political History of Britain 1870-1997*, 2nd ed., (London: Arnold Publishers, 1999).

R

Reese, Kenneth M., ed., *The American Chemical Society at 125: a Recent History, 1975-2001* (Washington, DC: ACS, 2002).

Rhead, M.M., "The Fate of DDT and PCBs in the Marine Environment," in *Specialist Periodical Reports: Environmental Chemistry Volume 1*, ed. G. Eglinton (London: CS, 1975), 137-159.

Riley, Dennis, William D. McGhee and Thomas Waldman, "Generation of Urethanes and Isocyanates from Amines and Carbon Dioxide," in *Benign by Design: Alternative Synthetic Design for Pollution Prevention*, eds. Paul T. Anastas and Carol A. Farris (Washington, DC: ACS, 1994), 122-132.

Rinia, Eduard Jan, "Measurement and Evaluation of Interdisciplinary Research and Knowledge Transfer," (PhD diss., Leiden University, 2007). In Leiden University Repository; available from https://openaccess.leidenuniv.nl/handle/1887/9923, accessed May 22, 2022.

Rip, Arie, "Onderzoek van de Ontwikkeling van Wetenschapsgebieden," *Kennis en Methode* 1 (1977): 311-333.

Rip, Arie, and Jan-Peter Voss, "Umbrella Terms as Mediators in the Governance of Emerging Science and Technology," *Science, Technology & Innovation Studies* 9 (2013): 39-59.

Roberts, Jody A., "Creating Green Chemistry: Discursive Strategies of a Scientific Movement," (PhD diss., Virginia Polytechnic Institute and State University, 2005). In VirginiaTechWorks; available from https://vtechworks.lib.vt.edu/handle/10919/27529, accessed May 22, 2022.

Rocke, Alan J., *Image & Reality: Kekulé, Kopp, and the Scientific Imagination* (London: The University of Chicago Press, 2010).

Roels, J.A., "Nu Slechts Veelbelovend, Straks Wellicht Realiteit," *Chemisch Magazine* 6/7 (1991): 339-341.

Rogers, Robin D., "First China-USA Green Chemistry Workshop," *Green Chemistry* 8 (2006): 126-127.

Rosenblum, Annette T., "Government Affairs," in *The American Chemical Society at 125: a Recent History, 1975-2001*, ed. Kenneth M. Reese (Washington, DC: ACS, 2002), 64-74.

Royston, Michael G., "Making Pollution Prevention Pay," *Harvard Business Review* 58 (1980): 6-27.

Ruckelshaus, William D., "Science, Risk and Public Policy," *Science* 26 (1983): 1026-1028.

Ruckelshaus, William D., "Environmental Protection: a Brief History of the Environmental Movement in America and the Implications Abroad," *Environmental Law* 15 (1985): 455-469.

S

Salmi, T., et al., "Development of Clean Technology for the Production of Triols," *Green Chemistry* 1 (1999): 283-287.

Sanderson, Katharine, "Chemistry: It's not Easy Being Green," *Nature* 469 (2011): 18-20.

Schummer, Joachim, "The Notion of Nature in Chemistry," *Studies in History and Philosophy of Science Part A* 34 (2003): 705-736.

Schummer, Joachim, "The Philosophy of Chemistry," *Endeavour* 27 (2003): 37-41.

Schwartz, Koos, "Milieuconvenant Stemt tot Tevredenheid," *Trouw*, January 18, 1996, accessed May 22, 2022, https://www.trouw.nl/home/milieuconvenant-stemt-tot-tevredenheid~a630f24a/.

Schwarz, Astrid E., "Green Dreams of Reason. Green Nanotechnology Between Visions of Excess and Control," *NanoEthics* 3 (2009): 109-118.

[SCP], *Sociaal en Cultureel Rapport 1998: 25 Jaar Sociale Verandering* (Rijswijk: SCP, 1998).

[Secretaries of State for Environment, Trade and Industry, Health, Education and Science, Scotland, Transport, Energy and Northern Ireland], the [Minister of Agriculture, Fisheries and Food] and the [Secretaries of State for Employment and Wales], *This Common Inheritance: Britain's Environmental Strategy* (London: HMSO Publications, 1990).

Shapin, Steven, "Discipline and Bounding: The History and Sociology of Science as Seen through the Externalism-internalism Debate," *History of Science* 30 (1992): 333-369.

Sheail, John, "'Burning Bings': a Study of Pollution Management in Mid-twentieth Century Britain," *Journal of Historical Geography* 31 (2005): 134-148.

Sheldon, Roger A., "Organic Synthesis – Past, Present and Future," *Chemistry and Industry* 23 (1992): 903-906.

Sheldon, Roger A., "Selective Catalytic Synthesis of Fine Chemicals: Opportunities and Trends," *Journal of Molecular Catalysis A: Chemical* 107 (1996): 75-83.

Sheldon, Roger, "Aiming for that Goal of Zero Emission," review of *Green Chemistry: Theory and Practice*, by Paul T. Anastas and John C. Warner, *Nature* 399, no. 6731 (1999): 33.

Sheldon, Roger, "Green Chemistry – One Year On," *Green Chemistry* 2 (2000): G1-G4.

Sheldon, Roger A., "The E Factor: Fifteen Years On," *Green Chemistry* 9 (2007): 1273-1283.

Sheldon, R.A., *E Factors, Green Chemistry and Catalysis: Records of the Travelling Chemist* (Delft: TU Delft, 2007).

Sheldon, Roger A., "Green and Sustainable Chemistry: Challenges and Perspectives," *Green Chemistry* 10 (2008): 359-360.

Sijm, Dick, "Nieuwe Uitdagingen in de Milieuchemie," in Sectie Milieuchemie, *Jaarboek 1998: Sectie Milieuchemie* (Den Haag: KNCV, 1999), 98-102.

Simmons, Milagros S., "The Role of Catalysts in Environmentally Benign Synthesis of Chemicals," in *Green Chemistry: Designing Chemistry for the Environment*, eds. Paul T. Anastas and Tracy C. Williamson (Washington, DC: ACS, 1996), 116-130.

Sjöström, Jesper, "Green Chemistry in Perspective – Models for GC Activities and GC Policy and Knowledge Areas," *Green Chemistry* 8 (2006): 130-137.

Sjöström, Jesper, "The Discourse of Chemistry (and Beyond)," *Hyle* 13 (2007): 83-97.

Sjöström, Jesper, Ingo Eilks and Vânia G. Zuin, "Towards Eco-reflexive Science Education: a Critical Reflection about Educational Implications of Green Chemistry," *Science & Education* 25 (2016): 321-341.

Slagmaat, Christian A.M.R. van, et al., "Solvent-free Hydrogenation of Levulinic Acid to γ-Valerolactone Using a Shvo Catalyst Precursor: Optimization, Thermodynamic Insights, and Life Cycle Assesment," *Green Chemistry* 22 (2020): 2443-2458.

Small, Henry G., "A Co-citation Model of a Scientific Specialty: A Longitudinal Study of Collagen Research," *Social Studies of Science* 7 (1977): 139-166.

Small, Henry G., "Cited Documents as Concept Symbols," *Social Studies of Science* 8 (1978): 327-340.

Smith, Keith, Ping He and Ashley Taylor, "Selective *Para*-bromination of Phenyl Acetate under the Control of Zeolites, Bases, Acetic Anhydride or Metal Acetates in the Liquid Phase," *Green Chemistry* 1 (1999): 35-38.

Smith, Keith, Guy M. Pollaud and Ian Matthews, "Selective Production of 1-Arylalkenes. Synthesis of 1-Aryl-2-alkenes and 1-Arylalkanes *via* Friedel-Crafts Alkylation of Aromatic Substrates with Allylic Alcohols over Solid Acid Catalysts," *Green Chemistry* 1 (1999): 75-81.

"Solid Waste Disposal Act," in *Compilation of Selected Acts within the Jurisdiction of the Committee on Commerce: Environmental Law*, [Committee on Commerce U.S. House of Representatives] (Washington, DC: US Government Printing Office, 1995), 455-613.

Soveroski, Marie, "EC Enlargement and the Development of European Environmental Policy: Parallel Histories, Divergent Paths?," *Review of European Community and International Environmental Law* 13 (2004): 127-134.

Spitz, Peter H., ed., *The Chemical Industry at the Millennium: Maturity, Restructuring, and Globalization* (Philadelphia, PA: Chemical Heritage Press, 2003).

Spitz, Peter H., "The Chemical Industry and the Environment: Meeting the Challenge," in *The Chemical Industry at the Millennium: Maturity, Restructuring, and Globalization*, ed. Peter H. Spitz (Philadelphia, PA: Chemical Heritage Press, 2003), 206-246.

"Steelworkers to Act on Air Pollution," *Chemical & Engineering News* 47 (1969): 22.

Steinhauser, Klaus G., Steffi Richter and Jutta Penning, "Sustainable Chemistry in Dessau – A Workshop Report," *Green Chemistry* 6 (2004): G41-G43.

Steinzor, Rena I., "Reinventing Environmental Regulation: the Dangerous Journey from Command to Self-control," *Harvard Environmental Law Review* 22 (1998): 103-202.

Stephan, David G., and John Atcheson, "The EPA's Approach to Pollution Prevention," *Chemical Engineering Progress* 85 (1989): 53-58.

Stern, Michael K., "Nucleophilic Aromatic Substitution for Hydrogen: New Halide-free Routes for Production of Aromatic Amines," in *Benign by Design: Alternative Synthetic Design for Pollution Prevention*, eds. Paul T. Anastas and Carol A. Farris (Washington, DC: ACS, 1994), 133-143.

Steverink-de Zoete, Marian C., et al., "Enzymatic Synthesis and NMR Studies," *Green Chemistry* 1 (1999): 153-156.

Stoneman, Paul, "Government Spending on Research and Development in the UK," *Fiscal Studies* 20 (1999): 223-259.

Stringer, Judy, "Green Chemistry – EPA Launches Challenge," *Chemical Week* 157 (1995): 31.

Sulmann, Gerda, "Duurzame Chemie Vereist een Sprong Vanúit de Toekomst," *Chemisch Weekblad* 91 (1995): 1.

Sundström, G., and O. Hutzinger, "Environmental Chemistry of Flame Retardants V^{x}. The Composition of Bromkal ® 70-5 DE-A Pentabromodiphenyl Ether Preparation," *Chemosphere* 5 (1976): 187-190.

"Superfund Amendments and Reauthorization Act of 1986," in *Compilation of Selected Acts within the Jurisdiction of the Committee on Commerce: Environmental Law*, [Committee on Commerce U.S. House of Representatives] (Washington, DC: US Government Printing Office, 1995), 985-1039.

T

Tanko, James M., Joseph F. Blackert and Mitra Sadeghipour, "Supercritical Carbon Dioxide as a Medium for Conducting Free-radical Reactions," in *Benign by Design: Alternative Synthetic Design for Pollution Prevention*, eds. Paul T. Anastas and Carol A. Farris (Washington, DC: ACS, 1994), 98-113.

Tarr, Joel A., "Industrial Waste Disposal in the United States as a Historical Problem," *Ambix* 49 (2002): 4-20.

[Taskforce KNCV], *Milieu: Chemici een Zorg* (Den Haag: KNCV, 1990).

Terlouw, J.K., J.J. de Ridder, W. Heerma and G. Dijkstra, "Mass-spectrometric Determination of Metal Chelates," *Fresenius' Zeitschrift für Analytische Chemie* 249 (1970): 296-301.

Trost, Barry M., "The Atom Economy – A Search for Synthetic Efficiency," *Science* 254 (1991): 1471-1477.

Tuinstra, Willemijn, Leen Hordijk and Carolien Kroeze, "Moving Boundaries in Transboundary Air Pollution Co-production of Science and Policy under the Convention on Long Range Transboundary Air Pollution," *Global Environmental Change* 16 (2006): 349-363.

Tundo, P., "Report of the 1st meeting of the Working party on Green and Sustainable Chemistry" (European Association for Chemical and Molecular Sciences, Rome, February 28, 2008).

Tundo, Pietro, et al., "Synthetic Pathways and Processes in Green Chemistry. Introductory Overview," *Pure and Applied Chemistry* 72 (2000): 1207-1228.

Tundo, Pietro, and Fabio Aricò, "Green Chemistry on the Rise: Thoughts on the Short History of the Field," *Chemistry International* 29 (2007): 4-7.

Tundo, Pietro, and Liliana Mammino, eds., *Green Chemistry in Africa* (Venice: IUPAC and INCA, 2002).

Tunzelmann, Nick von, "Technology in Post-war Britain," in *The Cambridge Economic History of Modern Britain: Structural Change and Growth, 1939-2000*, eds. Roderick Floud and Paul Johnson (Cambridge: Cambridge University Press, 2004), 299-331.

Turnhout, Esther, et al., "New Roles of Science in Society: Different Repertoires of Knowledge Brokering," *Science and Public Policy* 40 (2013): 354-365.

[Tweede Kamer der Staten-Generaal], *Milieuprogramma 2001-2004: Vergaderjaar 2000-2001* ('s-Gravenhage: SDU Uitgevers, 2000).

U

[UK Cabinet Office], *Realising our Potential: a Strategy for Science, Engineering and Technology* (London: HMSO Publications, 1993).

"UK Group to Explore Potential Benefits of 'Foresight' Exercise," *Nature* 373 (1995): 271.

[UNEP], *Global Chemicals Outlook II: From Legacies to Innovative Solutions: Implementing the 2030 Agenda for Sustainable Development* (United Nations Environment Programme, 2019).

[United Nations], *Report of the United Nations Conference on the Human Environment* (Stockholm: United Nations, 1972).

[United Nations], *Convention on Long-Range Transboundary Air Pollution* (Geneva: United Nations, 1979). In United Nations Treaty Collection; available from https://treaties.un.org/Pages/ViewDetails.aspx?src=TREATY&mtdsg_no=XXVII-1&chapter=27&clang=_en, accessed May 22, 2022.

[United Nations], *Montreal Protocol on Substances that Deplete the Ozone Layer* (Montreal: United Nations, 1987).

[United Nations], *United Nations Framework Convention on Climate Change* (Rio de Janeiro: United Nations, 1992).

[United Nations Economic Commission for Europe], *Updated Handbook for the 1979 Convention on Long-range Transboundary Air Pollution and its Protocols* (Geneva: United Nations, 2015).

[United Nations World Commission on Environment and Development], *Our Common Future* (Oxford: Oxford University Press, 1987).

Ure, A.M., and M.L. Berrow, "The Elemental Constituents of Soils," in *Specialist Periodical Reports: Environmental Chemistry Volume 2*, ed. H.J.M. Bowen (London: RSC, 1982), 94-204.

V

Varma, Rajender S., "Solvent-free Organic Syntheses Using Supported Reagents and Microwave Irradiation," *Green Chemistry* 1 (1999): 43-55.

Varma, Rajender S., Kasi Pitchumani and Kannan P. Naicker, "Triphasic Catalyst Systems Based on Surfactant/Clay Composites. Facile Synthesis of Cyano, Thiocyano and Hydroxy Compounds Using a Triphasic Catalyst," *Green Chemistry* 1 (1999): 95-97.

Velde, Henk te, "Het Poldermodel. Een Introductie," in *Harmonie in Holland: Het Poldermodel van 1500 tot Nu*, eds. Dennis Bos, Maurits Ebben and Henk te Velde (Amsterdam: Uitgeverij Bert Bakker, 2011), 9-29.

Vergragt, Philip J., and Leo Jansen, "Sustainable Technological Development: The Making of a Dutch Long-term Oriented Technology Programme," *Project Appraisal* 8 (1993): 134-140.

Verstegen, Simon W., "Stedelijke en Industriële Vervuiling," in *Groene Geschiedenis van Nederland*, eds. Jan L. van Zanden and Simon W. Verstegen (Utrecht: Het Spectrum, 1993), 117-150.

Victor, David G., et al., eds., *The Implementation and Effectiveness of International Environmental Commitments: Theory and Practice* (Cambridge, MA: the MIT Press, 1998).

Vollhardt, Peter, and Neil Schore, *Organic Chemistry: Structure and Function*, 7th ed., (New York, NY: W.H. Freeman and Company, 2003).

W

Warner, John C., Paul T. Anastas and Jean-Pierre Anselme, "The Wittig Reaction in the Undergraduate Organic Laboratory," *Journal of Chemical Education* 62 (1985): 346.

"Washington: Bad Choice, Worse Timing," *Time*, July 16, 1984, accessed May 22, 2022, http://content.time.com/time/magazine/article/0,9171,951225,00.html.

Wentzel, Bastiënne, "Chemicaliën uit Groenafval," *C_2W* 108 (2012): 20-23.

[Werkgroep Aspekten van Milieuproblematiek in de Chemische Opleiding], *De Milieuchemie in het Wetenschappelijk Onderwijs* (Den Haag: KNCV, 1976).

[Wetenschappelijke en Technologische Commissie voor de Biobased Economy], *Naar Groene Chemie en Groene Materialen: Kennis- en Innovatieagenda voor de Biobased Economy* (Den Haag: WTC, 2011).

"Wet Milieubeheer," *Staatsblad van het Koninkrijk der Nederlanden* 80 (1994): 1-98.

Whiffen, David H., and Donald H. Hey, *The Royal Society of Chemistry: The First 150 Years* (London: RSC, 1991).

Wilson, K., A. Rénson and J.H. Clark, "Novel Heterogeneous Zinc Triflate Catalysts for the Rearrangement of α-Pinene Oxide," *Catalysis Letters* 61 (1999): 51-55.

Wilson, Michael P., and Megan R. Schwarzman, "Green Chemistry: Wilson and Schwarzman Respond," *Environmental Health Perspectives* 117 (2009): A386.

Wilson, Michael P., and Megan R. Schwarzman, "Toward a New U.S. Chemicals Policy: Rebuilding the Foundation to Advance New Science, Green Chemistry, and Environmental Health," *Environmental Health Perspectives* 117 (2009): 1202-1209.

Winsemius, Pieter, "Opening Address," *The Science of the Total Environment* 45 (1985): 7-15.

Winterton, Neil, "Twelve More Green Chemistry Principles," *Green Chemistry* 3 (2001): G73-G75.

Woltjer, Jan J., *Recent Verleden: De Geschiedenis van Nederland in de Twintigste Eeuw* (Amsterdam: Uitgeverij Balans, 1992).

Woodhouse, Edward J., and Steve Breyman, "Green Chemistry as Social Movement?," *Science, Technology & Human Values* 30 (2005): 199-222.

Woodin, S.J., "Environmental Effects of Air Pollution in Britain," *Journal of Applied Ecology* 26 (1989): 749-761.

Wray, K. Brad, "Rethinking the Size of Scientific Specialties: Correcting Price's Estimate," *Scientometrics* 83 (2010): 471-476.

Z

Zanden, Jan L. van, "Omstreden Landbouw," in *Groene Geschiedenis van Nederland*, eds. Jan L. van Zanden and Simon W. Verstegen (Utrecht: Het Spectrum, 1993), 63-91.

Zanden, Jan L. van, "De Veranderende Relatie tussen Mens en Natuur," in *Groene Geschiedenis van Nederland*, eds. Jan L. van Zanden and Simon W. Verstegen (Utrecht: Het Spectrum, 1993), 179-208.

Zanden, Jan L. van, and Simon W. Verstegen, eds., *Groene Geschiedenis van Nederland* (Utrecht: Het Spectrum, 1993).

Zika, Rod G., and William J. Cooper, eds., *Photochemistry of Environmental Aquatic Systems* (Washington, DC: ACS, 1987).

Zitko, V., and O. Hutzinger, "Uptake of Chloro and Bromobiphenyls, Hexachloro and Hexabromobenzene by Fish," *Bulletin of Environmental Contamination and Toxicology* 16 (1976): 665-673.

Zon, Henk van, *Tachtig Jaar RIVM* (Assen: Van Gorcum, 1990).

Zon, H. van, "Geschiedenis van Milieuproblemen in Vogelvlucht," in *Milieuchemie. I. Onderzoeksmethoden in de Milieuchemie*, ed. M.A.M. Meester (Heerlen: Open Universiteit, 2000), 11-36.

(E) Important Websites

"Hands-on Class on Science & Technology Indicators," Loet Leydesdorff, accessed May 22, 2022, http://www.leydesdorff.net/indicators/lesson1.htm.

"The Nobel Prize in Chemistry 2005: Yves Chauvin, Robert H. Grubbs, Richard R. Schrock," The Royal Swedish Academy of Sciences, accessed May 22, 2022, http://www.nobelprize.org/nobel_prizes/chemistry/laureates/2005/press.html.

Mark Peplow, "Warning Shot for Green Chemistry," *Nature*, last modified November 3, 2005, http://www.nature.com/news/2005/051031/full/news051031-8.html.

Margaret Thatcher, "Speech to the Royal Society: September 27, 1988," Margaret Thatcher Foundation, accessed May 22, 2022, http://www.margaretthatcher.org/document/107346.

Yi Zhang, Bhavik R. Bakshi and Endalkachew Sahle-Demessie, "Evaluating the Greenness of Green Chemistry via Traditional and Thermodynamic Life Cycle Assessment," EPA, accessed May 22, 2022, https://cfpub.epa.gov/si/si_public_record_Report.cfm?dirEntryID=143141.

Samenvatting

In de jaren 90 van de twintigste eeuw ontstonden er binnen de chemische wetenschap nieuwe initiatieven onder de noemer van "groene chemie" die streefden naar meer milieuvriendelijke en duurzame chemische processen. Sindsdien maakt dat nieuwe werkterrein een spectaculaire groei door. Dit proefschrift, *Research between Science, Society and Politics: The History and Scientific Development of Green Chemistry*, probeert de drijfveren achter het ontstaan en de groei van groene chemie zichtbaar te maken en te verklaren.

Hiervoor kijk ik op het snijvlak van chemische wetenschap en samenleving naar de voorgeschiedenis, het ontstaan, en de institutionalisering van groene chemie. Daarbij is de aandacht in hoofdstuk 2 gericht op de Verenigde Staten (VS), in hoofdstuk 3 op het Verenigd Koninkrijk (VK) en, tenslotte, in hoofdstuk 4 op Nederland. Dit proefschrift laat zien dat de auteurs van de eerste invloedrijke boeken over groene chemie werkzaam waren in de VS en dat deze auteurs ook daarna dominant bleven in dit nieuwe veld. Het meest belangrijke wetenschappelijke tijdschrift voor de internationale groei van groene chemie – *Green Chemistry* – was echter van Britse makelij. Deze ontwikkelingen vormen een interessant vergelijkingsmateriaal voor de ontwikkelingen die op dit gebied in Nederland plaatsvonden. Bovendien biedt de vergelijking tussen die landen de mogelijkheid om meer zicht te krijgen op de algemene factoren die de opkomst van de groene chemie bevorderden, en de factoren die specifiek waren voor elk land afzonderlijk zodat uiteindelijk groene chemie beter kan worden gekarakteriseerd.

In hoofdstuk 1 laat ik zien dat de opkomst en de wetenschappelijke identiteit van groene chemie op verschillende – en soms tegengestelde – wijzen zijn gekarakteriseerd in de literatuur die tot nu toe over dit onderwerp is verschenen. In 2005 concludeerden Edward Woodhouse en Steve Breyman bijvoorbeeld dat de opkomst van groene chemie moet worden gezien als een 'sociale beweging', omdat die wordt gedreven door factoren buiten de muren van het laboratorium. In datzelfde jaar concludeerde Jody Roberts in zijn uitgebreide historische analyse *Creating Green Chemistry*, deels in reactie op Woodhouse en Breyman, dat groene chemie een 'wetenschappelijke beweging' is. Hij baseerde zich daarbij op de zogeheten "12 principes van de groene chemie" die in 1998 waren opgesteld door de Amerikanen Paul Anastas en John Warner. Deze principes bepalen, volgens Roberts, de identiteit van groene chemie als een interdisciplinair onderzoeksveld. Jesper Sjöström daarentegen laat álle beleids- en kennisgebieden die streven naar het "vergroenen" van de scheikunde vallen onder de noemer van groene chemie. Hij stelde in 2007 dat groene chemie is ontstaan in reactie op het door milieuproblemen veroorzaakte negatieve publieke imago van de chemie. Volgens hem is groene chemie een 'meta-discipline' die ontstond om sociaal-culturele redenen. Marianna Epicoco en haar coauteurs probeerden in 2014 verschillende van de hiervoor genoemde visies te integreren door te stellen dat 'groene chemici' binnen de chemische wetenschap een 'epistemische gemeenschap' vormen die, gebruikmakend van de 12 principes, probeert door middel van toegepast onderzoek specifieke vraagstukken uit de samenleving op te lossen. Weer anderen zien het gebruik van de term groene chemie als een poging tot 'greenwashing' om het publieke imago van de scheikunde op te poetsen. Zover gaat Bernadette Bensaude-Vincent niet in *Public Understanding of Science* (2014), maar volgens haar is de term groene chemie wel een "buzzword" dat vooral de retorische functie heeft om onbegrip tussen, in dit geval, de chemie en samenleving te reduceren.

Om deze zes visies op hun merites te beoordelen, én om een dieper inzicht in de aard van de groene chemie als vakgebied en retorisch repertoire te krijgen, maak ik gebruik van een aantal theorieën en concepten over de ontwikkeling van wetenschapsgebieden. Bijvoorbeeld van het model van Nicholas Mullins, dat ervan uitgaat dat nieuwe paradigma's een cruciale rol spelen bij de start van een wetenschappelijk specialisme. Verder gebruik ik het "parapluterm" concept van Arie Rip en Jan-Peter Voss. Zij beogen daarmee de opkomst en stabilisatie van nieuwe wetenschappelijke velden te karakteriseren op basis van zowel epistemische componenten alsook op basis van institutionalisering en het strategische gebruik van retorische instrumenten in relatie tot wetenschapsbeleid en tot zorgen die leven in de samenleving. Karakteristiek voor een parapluterm is dat deze zowel een inhoudelijke component heeft, die vaak ter discussie staat, alsook een retorische component. Tevens gebruik ik het model van Laurens Hessels en Harro van Lente dat zogeheten 'contracten tussen wetenschap en samenleving' op macroniveau beschrijft. Dit model gaat in op wat, en waarom, de wetenschap zou moeten doen en onder welke – vaak financiële – condities.

Met behulp van deze inzichten werpt de huidige studie een nieuw licht op de groei en identiteit van groene chemie. Een van de vragen waaraan ik in het bijzonder aandacht heb besteed is hoe milieuproblemen en het imago van de chemie het ontstaan en de ontwikkeling van groene chemie in de VS, het VK en Nederland hebben beïnvloed. Dit proefschrift onderzoekt deze maatschappelijke aspecten vooral op basis van een historische analyse van de betreffende ontwikkelingen in die drie landen, waarbij in het bijzonder wordt gekeken naar de rol en activiteiten van de nationale chemische verenigingen: de American Chemical Society (ACS), de Royal Society of Chemistry (RSC) en de Koninklijke Nederlandse Chemische

Vereniging (KNCV). Deze benadering leidt tot de identificatie van dominante actoren op het terrein van de groene chemie en tot meer inzicht in het retorisch gebruik van deze term in de dialoog tussen chemie en samenleving. De nationale ontwikkelingen worden besproken in de hoofdstukken 2, 3 en 4 in samenhang met de relevante politiek-culturele factoren uit de drie verschillende landen. Bij die analyse was Sheila Jasanoff's boek *Designs on Nature: Science and Democracy in Europe and the United States* zeer behulpzaam. Zij laat zien dat beleidsdoelen en de stijl van het wetenschapsbeleid sterk cultureel zijn bepaald, en onderscheidt daarbij drie typen politieke culturen: pluralisme, dat gepaard gaat met polarisatie; een op consensus gerichte politieke cultuur; en tenslotte een politieke cultuur waarin overheidsexperts een zeer actieve en dominante rol spelen. Deze indeling was ook voor mijn studie nuttig en ik neem die dan ook mee in mijn analyse van de ontstaansgeschiedenis van milieuchemie in de drie genoemde landen. Deze vormt tevens de voorgeschiedenis van de groene chemie.

In hoofdstuk 5 vindt vervolgens een conceptuele analyse plaats, in samenhang met een studie van chemische activiteiten die in het laboratorium plaatsvonden. Het doel hiervan is om groene chemie vanuit het perspectief van chemische grondslagen te begrijpen. Daarnaast kijk ik ook naar het retorisch gebruik van de term groene chemie. Het doel is om zo de balans te kunnen opmaken van de mate waarin het ontstaan van groene chemie in de jaren 90 verklaard kan worden door de rol van nieuwe chemische kennis, en de mate waarin louter retorische elementen daarbij een rol speelden. Hoofdstuk 6 voorziet in een comparatieve historische analyse waarin de resultaten van de hoofdstukken 2, 3, 4 en 5 worden besproken en met elkaar vergeleken om eindconclusies te trekken met betrekking tot het wetenschappelijk karakter en de drijfveren achter het ontstaan van groene chemie.

De hoofdstukken 2, 3 en 4 vormen het empirisch hart van dit boek. Hoofdstuk 2, over de VS, laat zien dat Amerikaanse chemici de eersten waren met diepgaande publicaties over groene chemie. Een dominante rol werd daarbij gespeeld door chemici werkzaam bij de US Environmental Protection Agency (EPA) en door een nieuwe milieuwet. Chemici van de EPA waren actief binnen de ACS Division of Environmental Chemistry (DEC). Deze milieuchemische sectie zag haar ledenaantal begin jaren 70 verdubbelen. In de jaren 80 zette de groei van dit sub-veld van de chemie verder door vanwege een toenemende vraag naar chemische kennis over het milieu. Er was in die jaren een sterke (juridische) polarisatie tussen de Amerikaanse overheid, de chemische industrie en milieu-gerelateerde ngo's. Tegelijk maakten de chemische industrie en de chemische wetenschap zich zorgen over het negatieve imago van de chemie. Mede vanwege de optredende polarisatie besloot de Amerikaanse overheid, in samenspraak met de industrie en andere partijen, de regelgeving aan te passen. Het Amerikaans Congres accordeerde de Pollution Prevention Act (PPA) in 1990. In plaats van het louter opleggen en handhaven van milieustandaarden stond deze wet juist voor de preventie van milieuvervuiling.

Deze wet bleek een keerpunt, want hij legde de basis voor groene chemie in de VS. Diverse chemici verbonden aan de EPA begonnen met PPA in de hand actief de boodschap van afvalpreventie uit te dragen en dan wel met name die van giftig afval dat ontstaat bij chemische processen. Vooral Paul Anastas en Joseph Breen die belast waren met beleidszaken verzorgden tientallen lezingen en symposia op nationale congressen van de ACS, vooral binnen DEC. Deze inspanningen leidden tot de publicatie van de invloedrijke symposiumbundels *Benign by Design: Alternative Synthetic Design for Pollution Prevention* (1994) en *Green Chemistry: Designing Chemistry for the Environment* (1996). Vooral laatstgenoemde publicatie zette de term "groene chemie" op de kaart. Deze nieuwe ontwikkeling viel min of meer samen met het overheidsprogramma *Reinventing Environmental Regulation* (1995) van de regering van President Bill Clinton en Vice-President Al Gore. Afvalpreventie kreeg daarbij een hoge prioriteit, veelal onder het label van groene chemie. *Reinventing Environmental Regulation* legde ook de basis voor de Presidential Green Chemistry Challenge Awards die in 1996 voor de eerste keer werden uitgereikt, om nieuwe milieuvriendelijke chemische processen te belonen die een voorbeeldfunctie zouden kunnen vervullen om het onderlinge vertrouwen tussen de chemische industrie, de chemie en de samenleving te vergroten. De organisatie hiervan was in handen van de EPA, in samenwerking met de ACS.

In 1997 volgde institutionalisering van groene chemie in de vorm van het Green Chemistry Institute (GCI). Met financiële steun van de overheid, wetenschap en chemische industrie ging deze kleine organisatie de boodschap van groene chemie uitdragen onder chemici en beleidsmakers, en in de maatschappij. De ACS zag grote mogelijkheden om via het GCI het publieke imago van de chemie te verbeteren. Het maakte dit nieuwe instituut in 2000 onderdeel van de eigen organisatie. De ACS zag groene chemie als een veelbelovend instrument in de dialoog met politici en de samenleving om de relevantie van de chemie duidelijk te maken en onderzoeksgelden los te krijgen voor de chemie. De kans om een internationaal wetenschappelijk tijdschrift op het terrein van de groene chemie op te richten liet de ACS echter aan zich voorbij gaan. Het VK had de primeur.

Het tijdschrift *Green Chemistry* werd in 1999 opgericht door de RSC. In het VK volgde het ontstaan van groene chemie historisch gezien gedeeltelijk een andere route dan in de VS. Hoofdstuk 3 van dit proefschrift onderzoekt die geschiedenis. Net als in de VS werden omstreeks 1970 ook in het VK nieuwe milieuwetten ingevoerd. Juridische en institutionele hervormingen van milieubeleid vonden echter tamelijk geleidelijk plaats. In vergelijking met andere Noord-Europese landen was de publieke aandacht voor het milieu in het VK geringer. Daarnaast waren milieuboetes relatief laag en was er op het niveau van milieuhandhaving een nauwe en weinig transparante samenwerking tussen industrieën en overheden. De vraag naar milieuchemische kennis nam wel gestaag toe in het VK. In 1972 werd binnen de Industrial Division van de Chemical Society, een voorloper van de RSC, de Environment Group opgericht. De nauwe banden met de Industrial Division bleven ook na de vorming van de RSC intact, toen het ledenaantal van de Environment Group wat terugliep. Maar vanaf de late jaren 80 schoot het ledenaantal van deze werkgroep sterk omhoog om vervolgens in 1994 zo groot te worden dat de groep formeel onafhankelijk kon worden van de Industrial Division, en als RSC Environmental Chemistry Group (ECG) verder kon gaan.

De groei in ledenaantal hing samen met de toegenomen publieke aandacht voor milieuvervuiling en volksgezondheid. Dat zag ook de regering van Margaret Thatcher, die aanvankelijk weinig ophad met strenge milieustandaarden omdat die slecht spoorden met haar economisch-industrieel beleid. Vanwege toegenomen druk vanuit de Europese Economische Gemeenschap, maar vermoedelijk ook om electorale redenen, veranderde haar conservatieve regering van richting. In 1990 publiceerde de regering het eerste milieu-gerelateerde witboek van het VK, *This Common Inheritance: Britain's Environmental Strategy*. Inzet was ook het transparanter maken van het milieubeleid en de milieuhandhaving. Wat bleef, was de bestaande sterke focus op het concurrentievermogen van de industrie. Die focus kwam ook terug in het nieuwe wetenschaps- en technologiebeleid dat medio jaren 90 werd ontwikkeld onder de benaming 'Technology Foresight' (TF), dat deels ontstond als respons op vergelijkbare ontwikkelingen in andere landen zoals Nederland. Dit beleidsprogramma voor toegepaste wetenschap was een novum in de Britse politieke cultuur, omdat anders dan in voorgaande decennia de dialoog over het wetenschapsbeleid breder werd georganiseerd dan binnen de gebruikelijke, kleinschalige overheidscommissies. De RSC raakte ook betrokken bij TF. Sommigen leden van de RSC waren voorstander van het bevorderen van toegepaste wetenschap, maar er waren ook leden die voor een verschraling van de fundamentele chemische wetenschap vreesden. Binnen die context van dit nieuwe overheidsbeleid ontstonden er binnen de RSC mogelijkheden om netwerken te vormen die konden inspelen op de veranderende vraag naar chemische kennis, waaronder de preventie van giftig afval dat ontstaat bij chemische processen. De oprichting in 1998 van het RSC Green Chemistry

Network (GCN) en van het internationale tijdschrift *Green Chemistry* het jaar daarop, moeten in dat licht worden gezien.

De GCN en *Green Chemistry* werden in het VK gecoördineerd vanuit York University, door chemici uit het laboratorium van professor James Clark, die onder de noemer "schone technologie" zich al enkele jaren richtte op de ontwikkeling van (industriële) chemische processen die minder (giftig) afval genereerden. Hij ruilde deze term eind jaren 90 in voor "groene chemie". Clark beschouwde, net als leiders van de RSC, groene chemie als een geschikt instrument in de dialoog met de samenleving, mede om het publieke imago van de chemie te verbeteren. Dat verklaart ook de weloverwogen keuze voor de titel van het tijdschrift dat zich internationaal zou ontwikkelen tot het meest belangrijke wetenschappelijke tijdschrift op dit nieuwe terrein: *Green Chemistry*. Met de oprichting van dat tijdschrift, dat zich ook profileerde als magazine, kregen chemici de kans om te laten zien aan de samenleving dat ze goed gemotiveerd waren om milieuproblemen veroorzaakt door chemische processen aan te pakken en te voorkomen. De start van *Green Chemistry* bleek daarnaast een keerpunt in het gebruik van de term groene chemie in de wetenschap. Dit gebruik nam internationaal in de jaren daarna een grote vlucht, vooral in chemische kringen. In het begin verscheen *Green Chemistry* zes keer per jaar, maar door de toenemende internationale aandacht verdubbelde de verschijningsfrequentie in 2004.

Hoofdstuk 4 onderzoekt de ontstaansgeschiedenis van groene chemie in Nederland, die wezenlijk verschilde van die in de VS en het VK. In Nederland kwam in de jaren 90, anders dan in de twee genoemde landen, de term "duurzame chemie" namelijk meer op de voorgrond te staan dan de term "groene chemie". Net als in de VS en het VK namen ook in Nederland in de jaren 60 en 70 de zorgen van burgers over milieuvervuiling toe. In het begin van de jaren 70 leidde dat tot de invoering van milieuwetten en hervormingen van de Rijksoverheid ter bescherming van de volksgezondheid. In die periode werd ook de Sectie Milieuchemie (MC) opgericht binnen de KNCV. De vraag naar chemische kennis over het milieu nam immers toe, bovendien waren chemici zelf gemotiveerd om bij te dragen aan de oplossing van milieuproblemen. Het ging daarbij onder andere om het in kaart brengen van de impact en verspreiding van toxische stoffen. Na de oliecrisis van 1973 maakte de aandacht voor het milieu tijdelijk plaats voor meer aandacht naar de toestand van de economie. In de jaren 80 keerde echter het tij en verbreedde de aandacht voor het milieu zich substantieel in vergelijking met andere landen. Relatief veel inwoners van Nederland werden donateur of lid van een milieu-gerelateerde ngo. De door de Rijksoverheid opgelegde milieustandaarden waren streng in vergelijking met andere Europese landen. Daarnaast had Nederland, anders dan de VS en het VK, een leidende rol in de multilaterale *Convention on Long-Range Transboundary Air Pollution* (*LRTAP*) uit 1979, die vanaf 1983 van kracht werd.

In plaats van de polarisatie die in de VS optrad, ontwikkelde het milieubeleid zich, in lijn met de Nederlandse politieke cultuur, op basis van een breed gedragen consensus onder belanghebbenden. Het in 1988 gepubliceerde rapport *Zorgen voor Morgen: Nationale Milieuverkenning 1985-2010* van het Rijksinstituut voor Volksgezondheid en Milieu (RIVM) is hier een voorbeeld van, maar ook de door de Rijksoverheid geïnitieerde milieuconvenanten zoals "Koolwaterstoffen 2000" (1989). Het maken van nieuw beleid was er op gericht om uiteenlopende belangen en zienswijzen van chemische industrieën en milieu-gerelateerde ngo's, met de hulp van actieve overheden, te bediscussiëren en te balanceren.

Deze sterk gedecentraliseerde stijl van beleid maken was, zoals Arie Rip en Barend van der Meulen in 1998 betoogden, in de jaren 90 ook van toepassing op het Nederlandse wetenschaps- en technologiebeleid. In deze periode nam de vraag naar toegepaste chemische kennis toe om de preventie van afval, een efficiënter gebruik van (bij voorkeur hernieuwbare) grondstoffen en een duurzame ontwikkeling te realiseren. Deze doelen beïnvloedden ook het chemisch onderzoek. Het programma Duurzame Technologische Ontwikkeling-Chemie (DTO-Chemie) is daarvan een illustratie. In dit beleidsprogramma van de Rijksoverheid waren tal van overheidsorganen, ngo's, wetenschappers en bedrijven betrokken. Medio jaren 90 leidde dat op basis van milieuaspecten en economische overwegingen tot de formulering van veelbelovende wetenschapsgebieden die R&D-subsidies zouden moeten ontvangen. Oud-Kamerlid en chemicus Leo Jansen en voormalig KNCV-voorzitter Jan Mulderink speelden daarin een leidende rol. De KNCV bleef binnen DTO-Chemie echter wat op de achtergrond. De Rijksoverheid stuurde deels wel mee via vijf ministeries die dit beleidsprogramma financierden. Net als in de VS en het VK speelde de wens om het, betrekkelijk negatieve, publieke imago van de chemie te verbeteren een rol bij chemici die in DTO-Chemie participeerden.

Onder de vlag van DTO-Chemie kwam de term "duurzame chemie" naar voren in verschillende publicaties. Daarnaast was deze term prominent aanwezig in de institutionele opvolger van DTO-Chemie, de in 1997 opgerichte Stichting Duurzame Chemie Ontwikkeling (DCO). De keuze voor de term duurzame chemie was vooral ingegeven door het feit dat duurzame ontwikkeling een expliciet beleidsdoel van de nationale overheid was. Een term die daar bij aansloot had de voorkeur. Als gevolg hiervan trad in de jaren 90 de term duurzame chemie in Nederland meer op de voorgrond dan de term "groene chemie" (die beginjaren 90 vooral werd gebruikt in een context van hernieuwbare grondstoffen).

In hoofdstuk 5 wordt met behulp van een conceptuele analyse van groene chemie in de drie genoemde landen nader ingegaan op de chemische grondslagen van dit veld. Er wordt gekeken naar het retorisch gebruik van de term groene chemie en onderzocht in hoeverre "groene chemici" bijeen worden gehouden door hetzelfde chemische paradigma. Daarnaast heb ik vooral de inhoud van het tijdschrift *Green Chemistry* en de daarin gepubliceerde artikelen onderzocht. Tevens heb ik de conceptualisering van groene en duurzame chemie door dominante actoren uit de VS, het VK en Nederland geanalyseerd.

Afgaande op de hoeveelheid publicaties en de weerklank daarvan, was Paul Anastas sinds het begin van de jaren 90 ongetwijfeld een van de meest productieve en invloedrijke groene chemici. Hij timmerde eerst aan de weg met *Benign by Design* (1994) en *Green Chemistry: Designing* (1996). Katalyse kreeg daarbij een prominente plek. Het concept "atoomeconomie" – dat erop gericht is zoveel mogelijk atomen uit de reactanten terug te laten komen in het gewenste product – kwam uiteraard ook naar voren. In feite probeerde hij met zijn 'groene chemie filosofie' laboratoriumchemici te overtuigen om een aantal regels in acht te nemen, want dat zou meer milieuvriendelijke en duurzame resultaten opleveren. Hij vond hiervoor vooral weerklank nadat hij samen met John Warner in 1998 het handboek *Green Chemistry: Theory and Practice* had gepubliceerd. Dat boek werd een van de meest geciteerde publicaties over groene chemie ooit, vooral vanwege de definitie van groene chemie die erin wordt gegeven en het conceptuele raamwerk dat daarvoor wordt gepresenteerd bestaande uit de zogeheten "12 principes". Groene chemie staat voor het gebruiken van een of meer van deze principes om de productie van afval en milieuvervuiling te voorkomen, door in te zetten op katalyse en atoomeconomie. Tevens vonden Anastas en Warner dat groene chemische processen bij voorkeur plaats moesten vinden op basis van hernieuwbare grondstoffen en voorts bij kamertemperatuur om de energieconsumptie te reduceren. De 12 principes zijn geformuleerd als vuistregels voor het werk in het laboratorium. Dat is de vorm waarin de 'groene chemie filosofie' van Anastas is uitgewerkt. Het karakter van deze principes is niet technisch-dwingend, in de vorm van bijvoorbeeld formules of harde normen, maar het zijn aanbevelingen om chemische processen duurzamer te maken. Anastas en Warner streefden ernaar om traditionele praktijken van de chemie bij te sturen, niet om ze radicaal te vervangen.

Iets soortgelijks zien we bij Clark die, naast hoofdredacteur van *Green Chemistry*, tevens een van de meest productieve groene chemici was. In zijn openingsartikel in *Green Chemistry* (1999), tevens zijn meest geciteerde artikel in dit tijdschrift, koppelde hij de term groene chemie aan de "ideale synthese". Deze moest volgens Clark academisch geloofwaardig en industrieel relevant zijn. In plaats van het formuleren van een nieuwe chemische theorie deed hij, net als Anastas en Warner, enkele aanbevelingen met betrekking tot veiligheid en het voorkomen van afval, om zo een 'groenere scheikunde' en een duurzame ontwikkeling na te streven. Het concept van ideale synthese illustreerde hij, anders dan Anastas, vooral met de door hem zélf ontwikkelde synthetische methodologieën, die vaak gebaseerd waren op katalyse en op zijn interesse voor toepassingen in de industrie.

Nederlandse chemici van het eerste uur die publiceerden in *Green Chemistry* waren Cees Okkerse en voormalig KNCV-voorzitter Herman van Bekkum, hoogleraar organische scheikunde in Delft. Aanvankelijk waren zij prominent betrokken bij beleidsprogramma's gericht op "duurzame chemie". Als onderdeel van DTO-Chemie publiceerden zij in *Duurzaamheid en Chemie: Een Bundel Essays* (1996) hun pleidooi voor het gebruik van biomassa en fotovoltaïsche energie. Okkerse en Van Bekkum berekenden dat vanaf 2040 fossiele grondstoffen niet meer nodig zouden zijn voor industriële productie en energieconsumptie. Dezelfde berekeningen publiceerden zij in 1999 weer, maar dit keer in *Green Chemistry*, zonder overigens een chemische definitie van de term duurzame of groene chemie te geven. Ook door DTO-Chemie werd de term duurzame chemie niet zozeer chemisch-inhoudelijk gedefinieerd, maar gekoppeld aan onderzoeks-domeinen die volgens DTO-Chemie streefden naar 'materiële duurzame ontwikkeling' en die op steun van wetenschap en industrie zouden moeten rekenen. Daarnaast waren er milieugroeperingen die, deels ook betrokken bij DTO-Chemie, vonden dat duurzame chemie stond voor 'werkstrategieën voor duurzaamheid'.

Het voorbeeld van Okkerse en Van Bekkum illustreert ook dat chemici die streefden naar meer milieuvriendelijke en duurzame chemische processen, een podium kregen in *Green Chemistry* zonder dat ze een nauwkeurig omschreven concept van groene chemie aanhingen. Hoofdstuk 5 laat zien dat vele auteurs, uit tal van landen, in het eerste jaar van *Green Chemistry* voortbouwden op totaal verschillende intellectuele bronnen. Heel verschillende onderzoeksproblemen kwamen in *Green Chemistry* aan bod. Veel chemici, met evident verschillende specialistische achtergronden, konden blijkbaar aanhaken bij groene chemie.

Vanaf de start van groene chemie in het begin van de jaren 90 waren er twee soorten gebruikers van de term groene chemie, die ik passieve en actieve gebruikers noem. Beide gebruikers hebben als overeenkomst dat ze streefden naar meer milieuvriendelijke en duurzame chemische processen. De passieve gebruikers hanteerden de term groene chemie voornamelijk terloops, zonder zelf de chemische betekenis ervan te specificeren of wezenlijk voort te bouwen op een bestaand concept van groene chemie. Ook verwijzen zij, vooral bij de opkomst van groene chemie in de jaren 90, vrijwel nooit naar het werk van invloedrijke groene chemici zoals Clark, Warner en Anastas.

Actieve gebruikers van de term groene chemie gaven, met Clark en Anastas als de meest prominente voorbeelden, wel conceptuele betekenis aan groene chemie, maar dat was meestal in een vrij algemene zin, zoals in de vorm van vuistregels en richtlijnen en niet in de vorm van een specifieke chemische theorie. Deze gebruikers probeerden vervolgens andere chemici te overtuigen van de waarde van hun specifieke concept.

Hoofdstuk 5 laat tot slot ook zien dat steeds meer chemici kritiek hadden op de inhoud van groene chemie. Deze chemici probeerden te kwantificeren hoe schoon, of duurzaam, chemische processen in feite waren die van het label groene chemie waren voorzien. Het valt niet mee om een nieuw chemisch proces te ontwikkelen dat volgens milieuwetenschappelijke methoden zoals "life cycle assessment" daadwerkelijk meer milieuvriendelijk is. Dit maakt de wetenschappelijke status van groene chemie kwetsbaar. Kort samengevat komt het erop neer dat de revolutionaire groei in het gebruik van de term groene chemie, die startte in de jaren 90, niet verankerd is in een gedeeld, fundamenteel nieuw (scheikundig) paradigma. Van een wetenschappelijke revolutie in de zin van Thomas Kuhn is geen sprake.

Dit proefschrift besluit in hoofdstuk 6 met een comparatieve analyse van de bepalende factoren uit zowel de chemische wetenschap als de samenleving op het ontstaan van groene chemie in de drie onderzochte landen. De relevante ontwikkelingen in de VS, het VK en Nederland worden in dit hoofdstuk systematisch met elkaar vergeleken om na te gaan wat de overeenkomsten en verschillen zijn geweest in de ontwikkeling van de groene chemie. Ik concludeer in dit hoofdstuk dat "groene chemie" als een parapluterm moet worden gezien.

De groeiende grootschalige industriële productie na de Tweede Wereldoorlog, zorgde, hand in hand met een toenemende democratisering, voor een sterker maatschappelijk milieubewustzijn. De opkomst van milieuchemie in de VS, het VK en Nederland moet vooral in dat licht worden bezien. Hoofdstuk 6 laat zien dat met name in Nederland dit vakgebied sterk opkwam, en dat dit samenviel met relatief grote zorgen omtrent het milieu onder de bevolking en binnen de politiek; veel sterker dan in het VK. Karakteristieke verschillen in politieke cultuur verklaren ook de verschillen in de ontstaansgeschiedenis van groene chemie in de VS, het VK en Nederland. De groene chemie ontstond in de VS, het VK en Nederland daardoor langs verschillende goeddeels politiek-cultureel bepaalde historische lijnen die duidelijk verband hielden met veranderend overheidsbeleid en sociale vraagstukken in deze landen.

In de VS bleek de milieuwet PPA een keerpunt, waardoor chemici van EPA de boodschap van afvalpreventie gingen verspreiden. Vervolgens werden in 1996 de Presidential Green Chemistry Challenge Awards ingesteld, nadat chemici van de EPA al enkele jaren deze boodschap hadden gepredikt binnen de ACS en dat dan vooral binnen DEC. In het VK en Nederland namen de milieuchemische secties van respectievelijk de RSC en de KNCV niet op die manier de leiding. De termen duurzame chemie en groene chemie werden vanaf midden jaren 90 in Nederland vooral gebruikt in een door de Rijksoverheid geïnitieerde nieuw wetenschapsbeleidsprogramma (DTO). Het was onderdeel van een sterk gedecentraliseerde beleidsproces en een breed gedragen consensus waarbij ook chemici waren betrokken. De milieuchemische sectie (MC) van de KNCV speelde daarin nauwelijks een rol. Ook in het VK speelde de milieuchemische sectie (ECG) van de RSC geen wezenlijke rol in de opkomst van groene chemie. In het VK kreeg eind jaren 90 een praktiserend synthetisch chemicus (Clark), met industriële en maatschappelijke interesse, het vertrouwen van de RSC om het tijdschrift *Green Chemistry* op te richten en te leiden. In feite was dit een respons van de RSC op nieuw wetenschapsbeleid (TF) dat sterk gericht was op industriële belangen.

Deze verschillende historische ontwikkelingen laten ook zien dat er sprake was van institutionaliseringen. Andere voorbeelden van institutionaliseringen betroffen GCI (in de VS) en GCN (in het VK). De institutionaliseringen droegen, in de zin van Rip en Voss, bij aan de stabilisatie van het veld van groene chemie en aan de spectaculaire groei ervan. Dat laatste kreeg ook vorm met de Presidential Green Chemistry Challenge Awards, maar vooral met het meest belangrijke wetenschappelijke tijdschrift voor de groene chemie: *Green Chemistry*.

De keuze voor de term groene chemie bleek binnen de ACS en RSC vooral een strategisch-retorische achtergrond te hebben, om in te spelen op een toenemend milieubewustzijn in de samenleving en een verslechterend imago van de chemie. Dit wordt geïllustreerd door de instelling van de Presidential Green Chemistry Challenge Awards, die overduidelijk bedoeld waren om de chemie in een positiever daglicht te zetten. Maar ook de keuze van de titel *Green Chemistry*, die volgens Clark vooral was ingegeven door het slechter wordende imago van de chemie, laat zien dat de term groene chemie een strategisch-retorische functie had ter bevordering van de legitimatie van de chemie in de samenleving.

Maar anders dan bij louter "greenwashing" of een "buzzword" had het label "groene chemie" ook epistemische waarde. Dat dubbele karakter van de groene chemie maakt het begrip bij uitstek tot een parapluterm. De chemische inhoud onder de noemer "groene chemie" stond voor meer duurzame en meer milieuvriendelijke chemische processen, al dan niet gestoeld op de vuistregels en chemische aanbevelingen van bijvoorbeeld Anastas en

Warner. Die inhoud kan alleen niet geformuleerd worden in termen van een gedeeld chemisch paradigma dat zorgde voor duidelijke demarcatie, zoals bijvoorbeeld Epicoco en coauteurs lijken te impliceren met hun begrip "epistemische gemeenschap". Want in tegenstelling tot het model van Mullins voor het ontstaan en de groei van een wetenschappelijk specialisme, startte de opkomst van groene chemie begin jaren 90 niet met de vorming van een chemisch paradigma dat ontwikkeld werd door dominante actoren die onderling samenwerkten aan hetzelfde onderzoeksprobleem in een laboratorium. De door Clark, Anastas, en Warner geformuleerde vuistregels en principes, zoals atoomeconomie, hebben echter wel een inhoud die aan de chemie zelf refereert. Deze set van regels had een open karakter waardoor ze, anders dan Roberts voorspelde, onvoldoende samenhang vertoonden om tot een paradigma te kunnen worden gerekend. Dat leidde soms ook tot kritiek.

De Britse chemicus Neil Winterton, bijvoorbeeld, beweerde in *Green Chemistry* (2001) dat de vuistregels van de 12 principes zo open waren dat het moeilijk was om er niet aan te voldoen. Enerzijds illustreert deze stelling van Winterton de open wetenschappelijke identiteit van groene chemie waardoor veel chemici konden aanhaken, maar anderzijds miskent deze stelling de epistemische waarde van de activiteiten in het laboratorium die plaatsvonden onder de noemer van groene chemie. Tal van chemici publiceerden nieuwe chemische syntheses als alternatief voor oudere methodes, met gunstige milieugevolgen. Het succes van de nieuwe chemische processen werd veelal gedemonstreerd op basis van gerealiseerde milieuwinst. Voor het vergelijken daarvan bestaan echter geen eenduidige milieunormen of gemeenschappelijke groene chemische kenmerken die zorgen voor intellectuele samenhang. Daardoor waren de chemische kennisroutes naar milieuwinst vanaf de jaren 90 niet alleen waardevol, maar vooral ook zeer divers, en dat allemaal onder een paraplu van groene chemie die steeds groter werd.

Curriculum Vitae

Arjan (Johan Alfredo) Linthorst was born in Ruinen on March 6, 1978. He finished his secondary school (HAVO) at De Groene Driehoek in Hoogeveen in 1996. He then studied chemistry at Windesheim University of Applied Sciences, which he finished in 2001 with a 2nd degree teacher training specialization. This was followed by a master's degree at the University of Groningen in 2004 within the physical organic chemistry research group of Professor Jan Engberts. As part of his education he published an article on the influence of long-tailed alcohols on the solubilization of cationic di-n-hexadecyldimethylammonium bromide bilayers, supervised by Jan Engberts and Jaap Klijn. In 2005, he also received a master's degree at Fontys University of Applied Sciences for his 1st degree teacher training in chemistry education.

Since 2003, he has been fulltime teaching chemistry at pre-university level in the Netherlands and on Curaçao. He is also active in the design and publishing of new experiments in chemical education journals. Currently, he is employed as a chemistry teacher at CSG Dingstede, at Meppel.

In 2005 he started his research on the present dissertation as an "external PhD candidate" at Maastricht University. He published several articles and was awarded the New Scholars Award (the Society for the History of Alchemy and Chemistry) and the Pieter Langerhuizen Stipendium (the Royal Holland Society of Sciences and Humanities). In 2009 he became fellow of the Descartes Centre for the History and Philosophy of the Sciences and the Humanities (Utrecht University).

Research Impact

The Regulations for Obtaining the Doctoral Degree: Maastricht University require a reflection on the (scientific) impact of the knowledge gathered during this PhD project, including the spin-offs for society.

The title of this PhD thesis *Research between Science, Society and Politics: The History and Scientific Development of Green Chemistry* already indicates that major attention is paid to society. Public environmental concerns with respect to the chemical sciences are discussed in particular, in order to assess the meaning and emergence of the umbrella term "green chemistry". The subject of this book therefore bears relevance to society, which matches my general personal interests as well.

Sharing Research Findings

Part of the research done for this PhD project has already been published and, subsequently, communicated to society. The latter is an important step in valorization.[851] As a PhD candidate, I communicated my first research findings by giving lectures at symposia for historians and philosophers of chemistry.[852] But, I also lectured for audiences consisting of non-specialists in, for instance, my place of birth, and I wrote some popular science articles in Dutch as well.[853] Although valorization was not my first reason for carrying out PhD research, I clearly have contributed to that. This is illustrated by my article published in *Foundations of Chemistry* in 2010 (which is the most cited one of this journal and also among the most downloaded ones (± 11,000 downloads) according to Springer).[854] This article discusses the

[851] Andreas Mitzschke, "Elusive Publics: Understanding Techno-scientific Controversy and Democratic Governance in the GM Crops Debate," (PhD diss., Maastricht University, 2018), 225-230. In UM Research Publications; available from https://cris.maastrichtuniversity.nl/en/publications/elusive-publics-understanding-techno-scientific-controversy-and-d, accessed May 22, 2022. Mitzschke discusses in his Valorization Addendum the positions that scientists can adopt in the valorization of their knowledge production.
[852] For instance: J.A. Linthorst, "Green Chemistry: Image, Environment and Chemical Societies" (2012 Summer Symposium of The International Society for the Philosophy of Chemistry, Leuven, Belgium, August 7-10, 2012); J.A. Linthorst, "The Scientific Development of Green Chemistry" (9th International Conference for the History of Chemistry, Uppsala, Sweden, August 21-24, 2013).
[853] J.A. Linthorst, "Het Nut in Vuur en Vlam: Magie der Chemie" (lecture given at the meeting of Nut & Genoegen, Ruinen, April 7, 2013). See also for example: Arjan Linthorst, "De Ongrijpbare Chemie," NEMO Kennislink, last modified November 16, 2010, https://www.nemokennislink.nl/publicaties/de-ongrijpbare-chemie/.
[854] Johan A. Linthorst, "An Overview: Origins and Development of Green Chemistry," *Foundations of Chemistry* 12 (2010): 55-68. See for facts and figures about my article: "Foundations of Chemistry,"

first research findings of my PhD journey, and has been translated and published in Chinese.[855] According to Google Scholar it has been cited by chemists, who often used my article to sketch the importance of combatting environmental problems, and historians and philosophers of chemistry who attempted to characterize the identity of green chemistry by building on my insights. Major international organizations in domains of chemistry education, science policy and environmental policy also cited my article, and used the historical development of green chemistry that I sketched in their policy making processes.[856] Obviously, the results I published in *Foundations of Chemistry* have been read and utilized by scientists and policy officers from rather different professional backgrounds. In this thesis I publish new research findings that might again be relevant to people inside and outside the academic circle of historians and philosophers of chemistry, and which I hopefully will further be disseminated with the aid of a publisher.

Trust and the Chemical Sciences

The subfield of environmental chemistry grew significantly within the chemical sciences between 1970 and 2000, as witnessed by the increase in relative membership numbers of this subfield within the American Chemical Society (ACS), the Royal Society of Chemistry (RSC) and the Royal Netherlands Chemical Society [Koninklijke Nederlandse Chemische Vereniging] (KNCV). There were also small differences, as that growth was culturally conditioned and thus varied from one political culture to another.[857] But, the ACS, RSC and KNCV do have in common that environmental chemistry became an established subfield, which has been part and parcel of the chemical sciences since the 1970s. In addition to that, environmental policy measures of national governments have often been based on knowledge of environmental chemistry in order to improve the quality of the environment for the benefit of society. This topic is discussed in the chapters 2 through 4 in this thesis. So, this study in

---

Springer, accessed May 22, 2022, https://link.springer.com/article/10.1007/s10698-009-9079-4. See also: Johan A. Linthorst, "The Image of Chemistry and Curriculum Changes," *Educacion Quimica* 23 (2012): 240-242.

[855] On request of Shuang Li (Chinese Academy of Sciences) this article was translated into Chinese: Shuang Li, "简论绿色化学的起源与发展," *Science & Culture Review* 14 (2017): 44-59.

[856] For example: Environment Directorate, *Sustainable Chemistry: Evidence on Innovation from Patent Data* (Paris: OECD, 2011); UNEP, *Global Chemicals Outlook II: From Legacies to Innovative Solutions: Implementing the 2030 Agenda for Sustainable Development* (United Nations Environment Programme, 2019).

[857] Cf., Sheila Jasanoff, *Designs on Nature: Science and Democracy in Europe and the United States* (Princeton, NJ: Princeton University Press, 2005).

the history of chemistry shows as a spin-off that many chemists have been willing to improve environmental quality. Although we still have environmental challenges in front of us, it is also clear that important intellectual solutions have to come from the chemical sciences. Therefore, many scholars have concluded that the chemical sciences deserve more trust, and thus better conditions, from society.[858]

Wasting Public Money?

I have also argued that many chemists felt distrusted by society. They experienced a negative public image. Although that might appear unfair when all advantages and disadvantages of chemical knowledge are balanced, the distrust triggered a major change in terminology within the chemical sciences. The umbrella term green chemistry emerged during the 1990s within academic circles, but also at the interface of science and society. An important example is the launching of the Presidential Green Chemistry Challenge Awards in the USA.[859]

Since the 1990s chemists are increasingly emphasizing economic and environmental benefits of their discipline, not in the last place by calling it "green". Many chemists thus attempted to show the relevance of chemical knowledge to society and to improve the public image of the chemical sciences in order to receive better conditions for their research. They tried to raise the legitimacy and funding of their discipline. The chapters 2 through 4 also show that the term green chemistry developed along different culturally conditioned historical lines. Chapter 5 demonstrates that the term green chemistry lacks a unified internal scientific meaning. The precise content of green chemistry has been a matter of dispute at several occasions.

The social value of my research results presented in the two paragraphs above is twofold. Firstly, I demonstrated that time, public money and other resources have been spent in different countries to water down the negative public image of chemistry (as felt by chemists).[860] Secondly, I also argued that this went hand in hand with a major change in terminology used by chemists in the presentation of their (publicly funded) knowledge production. But, as I have shown, the precise meaning of that knowledge remains vague for

---

858 Laurens K. Hessels and Harro van Lente, "Practical Applications as a Source of Credibility: a Comparison of Three Fields of Dutch Academic Chemistry," *Minerva* 49 (2011): 218.
859 See Chapter 2 and 5 for the Presidential Green Chemistry Challenge Awards. See also: Arie Rip and Jan-Peter Voss, "Umbrella Terms as Mediators in the Governance of Emerging Science and Technology," *Science, Technology & Innovation Studies* 9 (2013): 39-59.
860 Of course, industries also spent money to green chemistry.

society. One can therefore raise the question whether public money has been wasted. By reading my thesis both society and the chemical sciences can get a stronger awareness of the roles played by umbrella terms such as green chemistry. Do they benefit us all?

Improving the Social Contract between Science and Society

The insights presented on the history of environmental chemistry, the emergence of green chemistry and the public image of chemistry, mentioned above, will hopefully be utilized by professionals in the policy domains of science, environment and education. In my view the most important opportunity is organizing a broadly conceived dialogue, starting in the Netherlands.

People pay taxes. A part of those taxes is spent on academic research. In order to carry out that research, scientists thus have to receive funding from society. Next to some "academic freedom", they have responsibility to society. Science and society therefore do have a relationship that can better be built on a reasonable basis of trust instead of distrust. In case of the chemical sciences, society should know that many chemists are willing to improve the environment, but meanwhile they feel often disrespected as well by that same society. The communication of chemical knowledge at the interface of science and society became therefore loaded with "umbrella terms" in the processes of obtaining research funding.[861]

These terms facilitate communication, but at the time – as I have shown – make it more obscure to the public at large which content is covered by those terms. That lack of understanding is influenced both by the negative public image of chemistry and by the poor scientific literacy of large parts of society.[862] It is important to change the tide. This thesis can trigger the start of a broadly conceived dialogue among stakeholders in our democracy – chemists from the laboratory included. That should lead to a covenant between NGOs, industries, government and academia that breathes an improved trust in science with a reduced place for (time and public money consuming) umbrella terms and give a bigger place for scientific content. Eventually, this could be followed by a reform of the public funding

[861] Most likely, that rhetorical behavior of scientist is enforced by structural changes in public funding mechanisms, which increasingly rely on strong competition among scientists. See: Mitzschke, "Elusive Publics: Understanding Techno-scientific Controversy and Democratic Governance in the GM Crops Debate", 225-230.

[862] See for the lack of scientific literacy: "PISA 2018 Science Framework," OECD, accessed May 22, 2022, https://www.oecd-ilibrary.org/sites/f30da688-en/index.html?itemId=/content/component/f30da688-en.

mechanisms of academic research. The time and public money saved could then be spent in the academic laboratory in order to produce new chemical knowledge, for the benefit of us all. That's what I would call an improved social contract between science and society.

Made in the USA
Monee, IL
05 November 2026

67807086R00157